The Student's
Anatomy
of Exercise Manual

The Student's
Anatomy
of Exercise Manual

Chief Consultant
Professor Ken Ashwell B.Med.Sc., M.B., B.S., Ph.D.

First edition for the United States, its territories and dependencies, and Canada published in 2012 by Barron's Educational Series, Inc.

First published in 2012 by Global Book Publishing Pty Ltd
181 Botany Road, Waterloo
NSW 2017, Australia
Email: rightsmanager@globalbookpublishing.com.au
www.globalbookpublishing.com.au

All inquiries should be addressed to:
Barron's Educational Series, Inc.
250 Wireless Boulevard
Hauppauge, NY 11788.
www.barronseduc.com

ISBN 978-1-4380-0113-5
Library of Congress Catalog Number: 2011938986

Printed in China by 1010 Printing International Ltd

9 8 7 6 5 4 3 2 1

It is recommended that anyone who is considering participating in an exercise program should consult a physician before starting and that no one should attempt a new exercise without the supervision of a certified professional. While every care has been taken in presenting this material, the anatomical and medical information is not intended to replace professional medical advice; it should not be used as a guide for self-treatment or self-diagnosis. Neither the authors nor the publisher may be held responsible for any type of damage or harm caused by the use or misuse of information in this book.

Publisher	James Mills-Hicks
Managing Editor	Barbara McClenahan
Project Manager	Selena Quintrell
Editor	Jennifer Taylor
Chief Consultant	Ken Ashwell B.Med.Sc., M.B., B.S., Ph.D.
Authors	Ken Ashwell B.Med.Sc., M.B., B.S., Ph.D.
	Michael Baker B.App.Sc., M.App.Sc., Ph.D., A.E.P.
	Tim Foulcher B.App.Sc., M.Phty.
	Michael Newton B.App.Sc., M.Sc., Ph.D., A.E.P.
Cover Design	Stan Lamond
Designer	Maria Harding
Design Concept	Stan Lamond
Illustrator (Exercises)	Kristen W. Marzejon M.A.M.S., C.M.I.
Illustrations Editor	Selena Quintrell
Indexer	Puddingburn Publishing Services
Proofreader	Amanda Burdon
Production Manager	Karen Young
Editorial Coordinator	Kristen Donath
Other Illustrations	David Carroll, Peter Child, Deborah Clarke, Geoff Cook, Marcus Cremonese, Beth Croce, Hans De Haas, Wendy de Paauw, Levant Efe, Mike Golding, Mike Gorman, Jeff Lang, Alex Lavroff, Ulrich Lehmann, Ruth Lindsay, Richard McKenna, Annabel Milne, Tony Pyrzakowski, Oliver Rennert, Caroline Rodrigues, Otto Schmidinger, Bob Seal, Vicky Short, Graeme Tavendale, Thomson Digital, Jonathan Tidball, Paul Tresnan, Valentin Varetsa, Glen Vause, Spike Wademan, Trevor Weekes, Paul Williams, David Wood

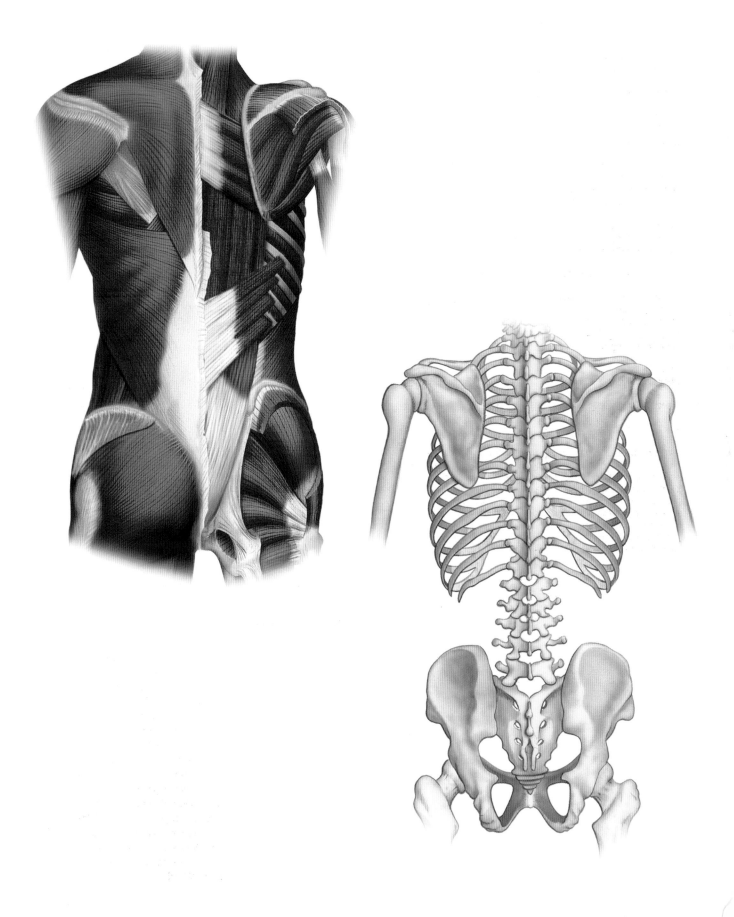

Contents

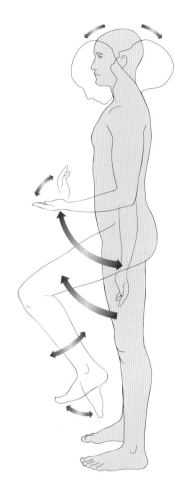

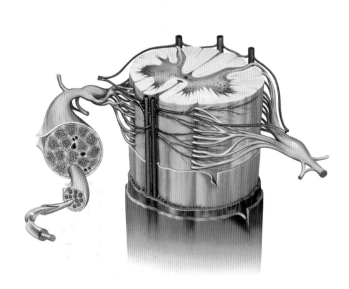

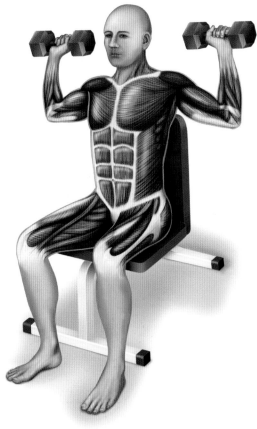

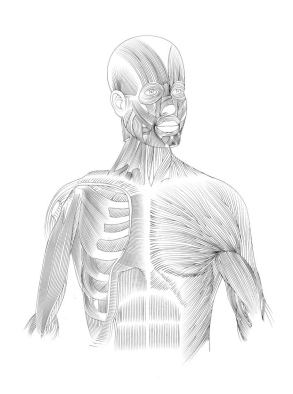

Foreword

There is a revolution in thinking about exercise. Exercise is no longer the sole domain of the sportsperson or body builder. Health professionals now know that physical activity is central to maintaining health and coordination for everyone from childhood to old age. Exercise is not a spectator sport and just buying a gym membership is not enough to keep your body in optimal shape. All adults should engage in an age-appropriate regime of exercise that takes their medical history and body type into account.

For the student of anatomy or sports physiology—who will train the athletes of tomorrow, as well as keeping all of the rest of us in shape—and the sportsperson, body builder, or just anyone wanting to improve their physical fitness, this book brings together sound practical advice on how to perform the most important exercises, with the anatomical detail to explain what each exercise is doing for the body. You will learn how specific exercises target particular muscle groups to ensure that you or your clients get the best result, whether for sporting needs or general fitness. The book includes an overview section at the front to explain the anatomy and function of key muscles, as well as a coloring workbook at the back to reinforce what you have learned.

Always remember to follow the advice and warnings for each exercise. Anyone beginning an exercise program should first check with their medical practitioner, particularly if they are over 40 years old or have a history of heart disease, or high blood pressure. An exercise professional at your local gym will advise on the combination of exercises that are best for you. Some exercises use heavy weights and test the strength and flexibility of muscles that we do not often use in our daily lives. "Do it right" means preparing properly for each exercise, using only the correct equipment, seeking help from an exercise professional or spotter where necessary, and considering the safety of those around you when performing the exercise. Safety for yourself and others should always be the highest priority.

Professor Ken Ashwell
Department of Anatomy,
School of Medical Sciences, Faculty of Medicine,
University of New South Wales, Sydney, Australia

How This Book Works

This book is organized into three primary sections: a full-color anatomy overview; a full-color illustrated exercise guide, comprising the main part of the book; and a coloring workbook in which to test your anatomical knowledge.

The anatomy overview section provides detailed, anatomically correct illustrations with clear, informative labels for the various body systems and regions. Visualizing the parts of the body and their links to each other will improve your understanding of how the body works during exercise.

Each of the five chapters in the exercise guide focuses on specific muscle areas—arms and shoulders, chest, back, trunk, and legs and buttocks. Every exercise is depicted with two anatomically correct poses. Labels identify all the important muscles—including identifying active and stabilizer muscles—so you can visualize and understand exactly which muscles are activated during the exercises. This will not only increase your knowledge of anatomy; it will help to improve the effectiveness of workouts and rehabilitation programs.

The coloring workbook chapter is a study aid that aims to facilitate your understanding of important body systems—the muscular, skeletal, and nervous systems. Color in each illustration to help memorize the location of muscles, bones, and nerves within these systems. Fill in the blank labels to test your knowledge of the names of body parts—the answers are given at the bottom of each page.

ANATOMY OVERVIEW PAGES

This section contains full-color, double-page overview spreads that give a rundown on the important parts of a particular body system.

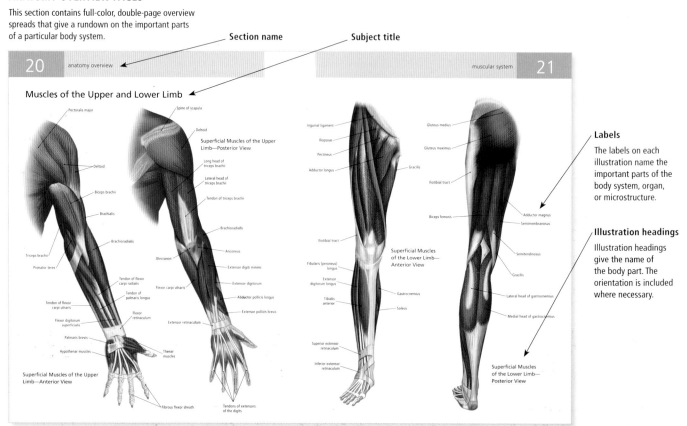

Section name

Subject title

20   anatomy overview

muscular system   21

Muscles of the Upper and Lower Limb

Labels

The labels on each illustration name the important parts of the body system, organ, or microstructure.

Illustration headings

Illustration headings give the name of the body part. The orientation is included where necessary.

EXERCISE PAGES

Each chapter in the exercise guide focuses on specific muscle areas. The exercises depict two anatomically correct poses that identify the active and stabilizer muscles.

Chapter name

Exercise title

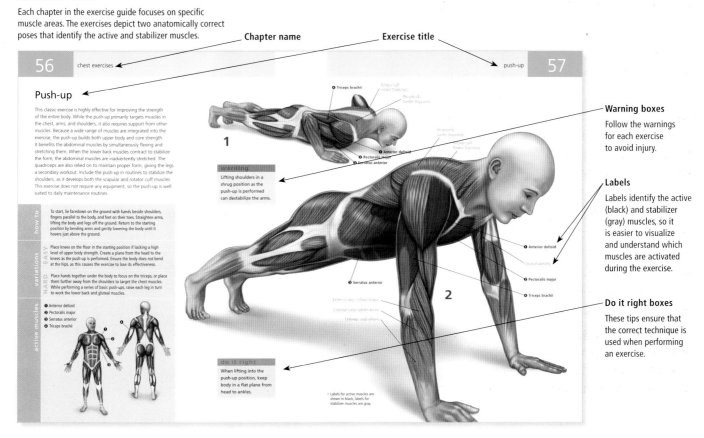

56 chest exercises

push-up 57

Push-up

This classic exercise is highly effective for improving the strength of the entire body. While the push-up primarily targets muscles in the chest, arms, and shoulders, it also requires support from other muscles. Because a wide range of muscles are integrated into the exercise, the push-up builds both upper body and core strength. It benefits the abdominal muscles by simultaneously flexing and stretching them. When the lower back muscles contract to stabilize the form, the abdominal muscles are inadvertently stretched. The quadriceps are also relied on to maintain proper form, giving the legs a secondary workout. Include the push-up in routines to stabilize the shoulders, as it develops both the scapular and rotator cuff muscles. This exercise does not require any equipment, so the push-up is well suited to daily maintenance routines.

how to
To start, lie facedown on the ground with hands beside shoulders, fingers parallel to the body, and feet on their toes. Straighten arms, lifting the body and legs off the ground. Return to the starting position by bending arms and gently lowering the body until it hovers just above the ground.

EASY Place knees on the floor in the starting position if lacking a high level of upper body strength. Create a plane from the head to the knees as the push-up is performed. Ensure the body does not bend at the hips, as this causes the exercise to lose its effectiveness.

HARD Place hands together under the body to focus on the triceps, or place them further away from the shoulders to target the chest muscles. While performing a series of basic push-ups, raise each leg in turn to work the lower back and gluteal muscles.

active muscles
❶ Anterior deltoid
❷ Pectoralis major
❸ Serratus anterior
❹ Triceps brachii

warning
Lifting shoulders in a shrug position as the push-up is performed can destabilize the arms.

do it right
When lifting into the push-up position, keep body in a flat plane from head to ankles.

❻ Triceps brachii
Rotator cuff (teres minor)
Rhomboids (under trapezius)

❹ Anterior deltoid
❷ Pectoralis major
❸ Serratus anterior

❶ Anterior deltoid
Latissimus dorsi
❷ Pectoralis major
❸ Triceps brachii

❸ Serratus anterior

Extensor carpi radialis longus
Extensor carpi radialis brevis
Extensor carpi ulnaris

1

2

* Labels for active muscles are shown in black, labels for stabilizer muscles are gray.

Warning boxes
Follow the warnings for each exercise to avoid injury.

Labels
Labels identify the active (black) and stabilizer (gray) muscles, so it is easier to visualize and understand which muscles are activated during the exercise.

Do it right boxes
These tips ensure that the correct technique is used when performing an exercise.

COLORING WORKBOOK PAGES

This final section contains black-and-white drawings of parts of the muscular, skeletal, and nervous systems. Color in the body parts as a memory aid.

Section name

Subject title

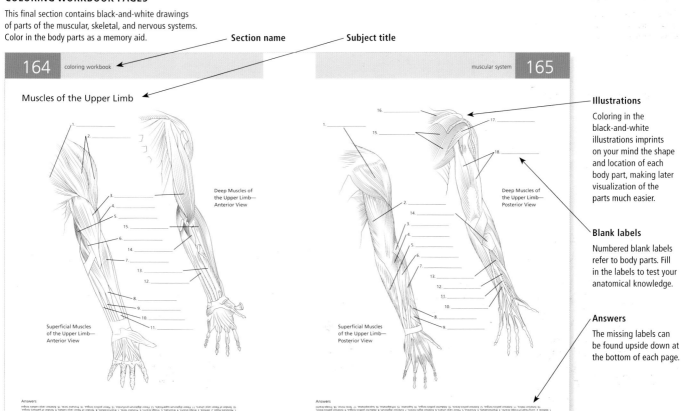

164 coloring workbook

muscular system 165

Muscles of the Upper Limb

Deep Muscles of the Upper Limb— Anterior View

Superficial Muscles of the Upper Limb— Anterior View

Deep Muscles of the Upper Limb— Posterior View

Superficial Muscles of the Upper Limb— Posterior View

Answers

Answers

Illustrations
Coloring in the black-and-white illustrations imprints on your mind the shape and location of each body part, making later visualization of the parts much easier.

Blank labels
Numbered blank labels refer to body parts. Fill in the labels to test your anatomical knowledge.

Answers
The missing labels can be found upside down at the bottom of each page.

Anatomy Overview

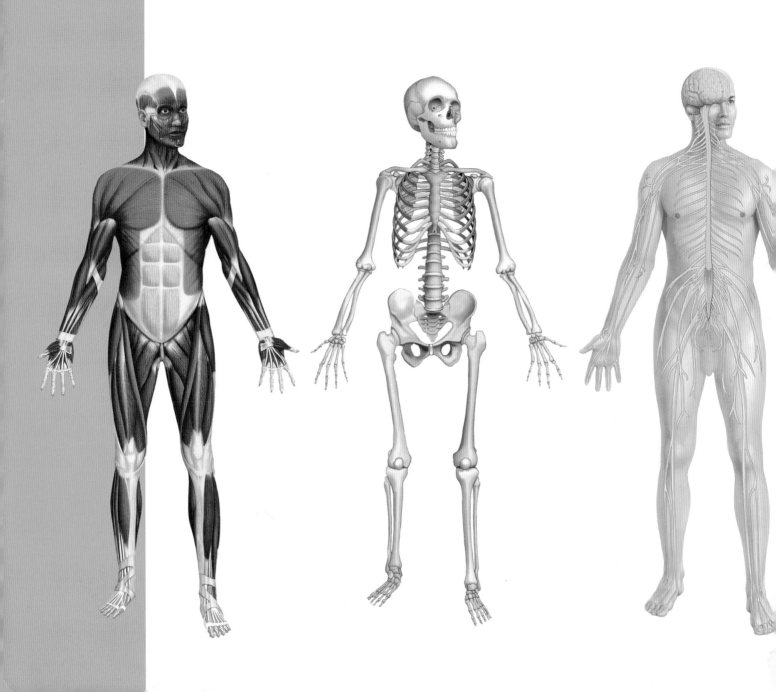

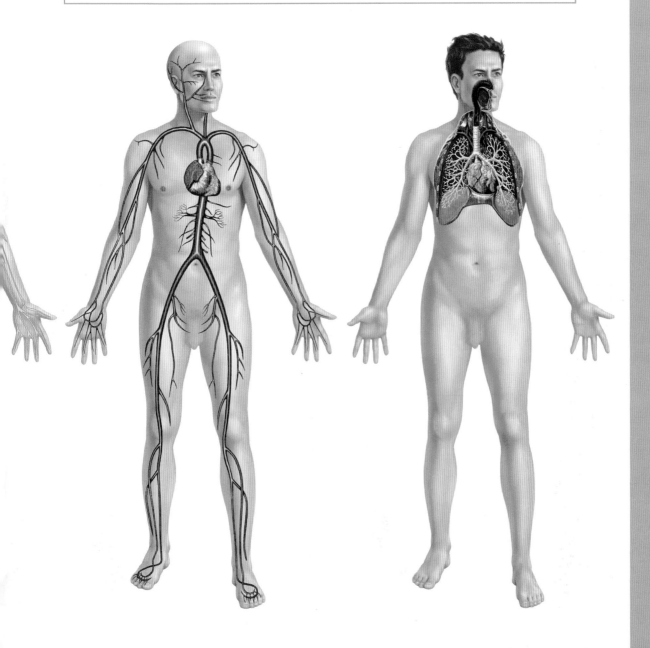

Body Regions

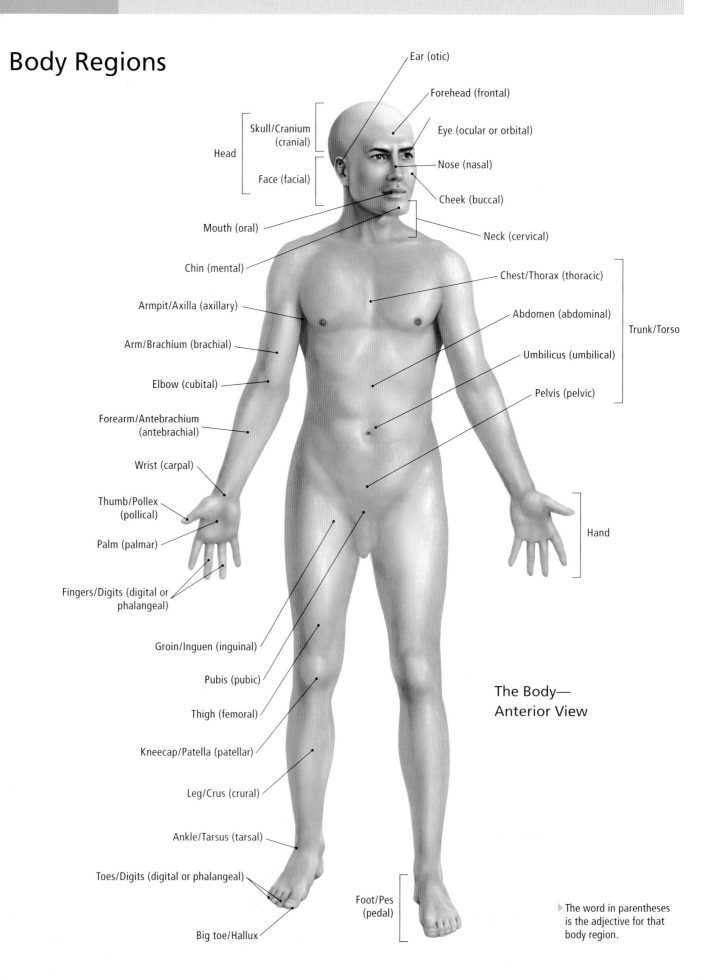

Ear (otic)

Forehead (frontal)

Skull/Cranium (cranial)

Eye (ocular or orbital)

Head

Nose (nasal)

Face (facial)

Cheek (buccal)

Mouth (oral)

Neck (cervical)

Chin (mental)

Chest/Thorax (thoracic)

Armpit/Axilla (axillary)

Abdomen (abdominal)

Arm/Brachium (brachial)

Umbilicus (umbilical)

Trunk/Torso

Elbow (cubital)

Pelvis (pelvic)

Forearm/Antebrachium (antebrachial)

Wrist (carpal)

Thumb/Pollex (pollical)

Hand

Palm (palmar)

Fingers/Digits (digital or phalangeal)

Groin/Inguen (inguinal)

Pubis (pubic)

Thigh (femoral)

The Body—
Anterior View

Kneecap/Patella (patellar)

Leg/Crus (crural)

Ankle/Tarsus (tarsal)

Toes/Digits (digital or phalangeal)

Foot/Pes (pedal)

▶ The word in parentheses
is the adjective for that
body region.

Big toe/Hallux

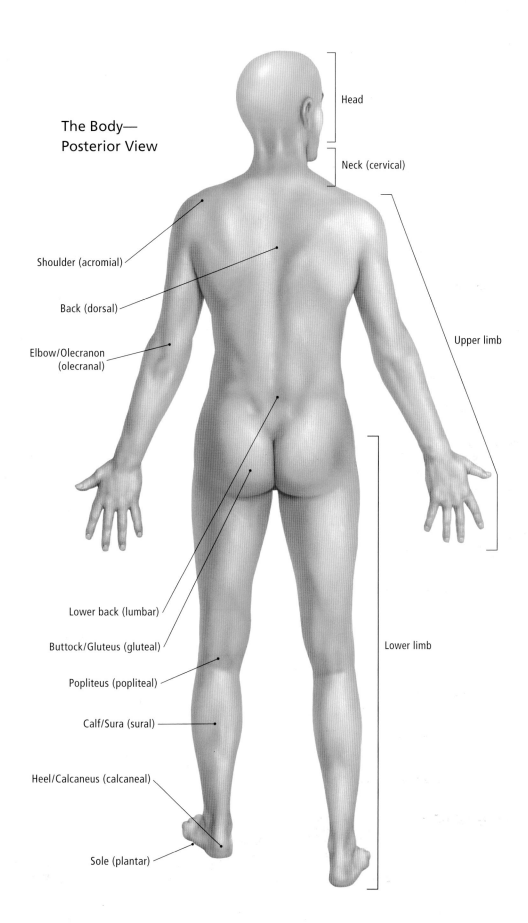

The Body—
Posterior View

Head

Neck (cervical)

Shoulder (acromial)

Back (dorsal)

Elbow/Olecranon
(olecranal)

Upper limb

Lower back (lumbar)

Buttock/Gluteus (gluteal)

Popliteus (popliteal)

Lower limb

Calf/Sura (sural)

Heel/Calcaneus (calcaneal)

Sole (plantar)

Muscles of the Body

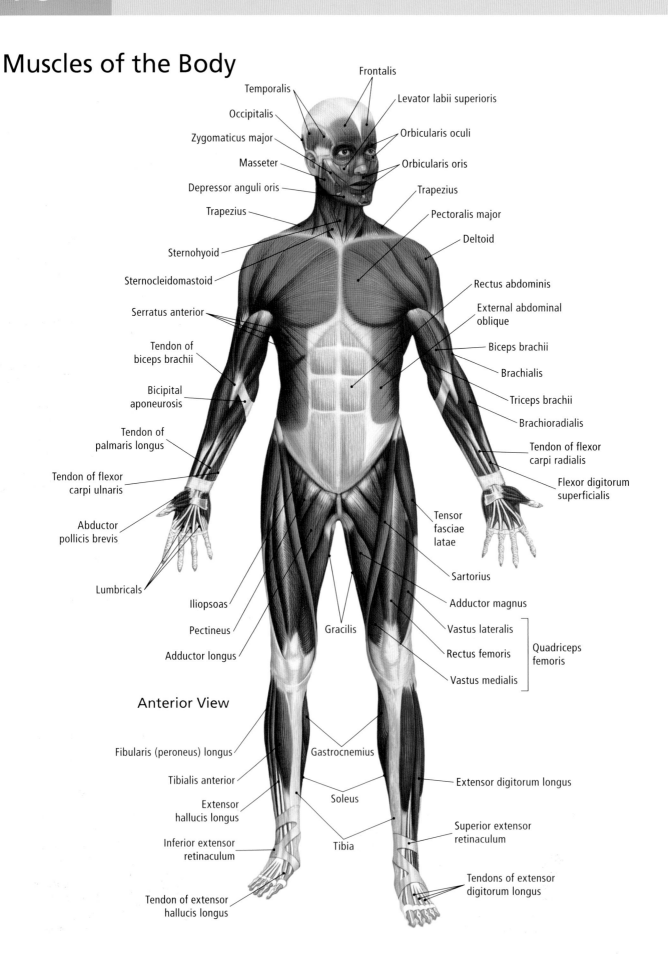

Frontalis

Temporalis

Levator labii superioris

Occipitalis

Orbicularis oculi

Zygomaticus major

Orbicularis oris

Masseter

Depressor anguli oris

Trapezius

Trapezius

Pectoralis major

Sternohyoid

Deltoid

Sternocleidomastoid

Rectus abdominis

Serratus anterior

External abdominal oblique

Tendon of biceps brachii

Biceps brachii

Brachialis

Bicipital aponeurosis

Triceps brachii

Tendon of palmaris longus

Brachioradialis

Tendon of flexor carpi radialis

Tendon of flexor carpi ulnaris

Flexor digitorum superficialis

Abductor pollicis brevis

Lumbricals

Tensor fasciae latae

Iliopsoas

Sartorius

Pectineus

Adductor magnus

Gracilis

Vastus lateralis

Adductor longus

Rectus femoris

Quadriceps femoris

Vastus medialis

Anterior View

Fibularis (peroneus) longus

Gastrocnemius

Tibialis anterior

Extensor digitorum longus

Extensor hallucis longus

Soleus

Superior extensor retinaculum

Inferior extensor retinaculum

Tibia

Tendon of extensor hallucis longus

Tendons of extensor digitorum longus

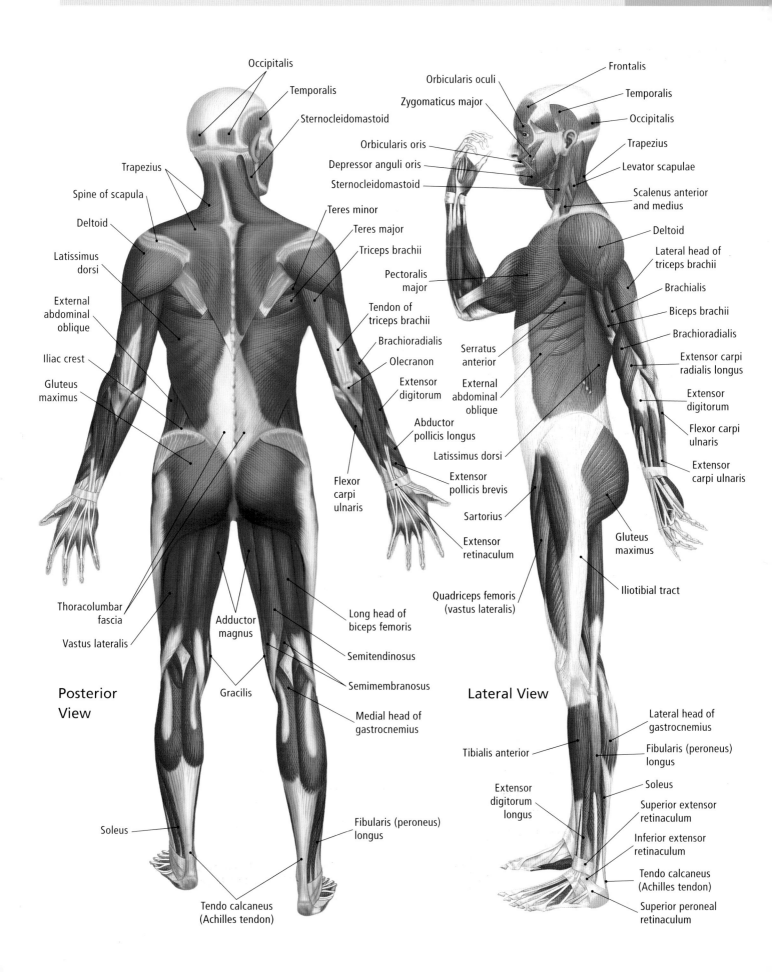

Occipitalis

Temporalis

Sternocleidomastoid

Trapezius

Spine of scapula

Deltoid

Latissimus dorsi

External abdominal oblique

Iliac crest

Gluteus maximus

Teres minor

Teres major

Triceps brachii

Pectoralis major

Tendon of triceps brachii

Brachioradialis

Olecranon

Extensor digitorum

Abductor pollicis longus

Flexor carpi ulnaris

Extensor pollicis brevis

Extensor retinaculum

Thoracolumbar fascia

Vastus lateralis

Adductor magnus

Gracilis

Long head of biceps femoris

Semitendinosus

Semimembranosus

Medial head of gastrocnemius

Soleus

Fibularis (peroneus) longus

Posterior View

Tendo calcaneus (Achilles tendon)

Orbicularis oculi

Zygomaticus major

Orbicularis oris

Depressor anguli oris

Sternocleidomastoid

Frontalis

Temporalis

Occipitalis

Trapezius

Levator scapulae

Scalenus anterior and medius

Deltoid

Lateral head of triceps brachii

Brachialis

Biceps brachii

Brachioradialis

Extensor carpi radialis longus

Extensor digitorum

Flexor carpi ulnaris

Extensor carpi ulnaris

Serratus anterior

External abdominal oblique

Latissimus dorsi

Sartorius

Gluteus maximus

Iliotibial tract

Quadriceps femoris (vastus lateralis)

Lateral View

Tibialis anterior

Extensor digitorum longus

Lateral head of gastrocnemius

Fibularis (peroneus) longus

Soleus

Superior extensor retinaculum

Inferior extensor retinaculum

Tendo calcaneus (Achilles tendon)

Superior peroneal retinaculum

Muscles of the Abdomen and Back

Muscles of the Abdomen—
Anterior View

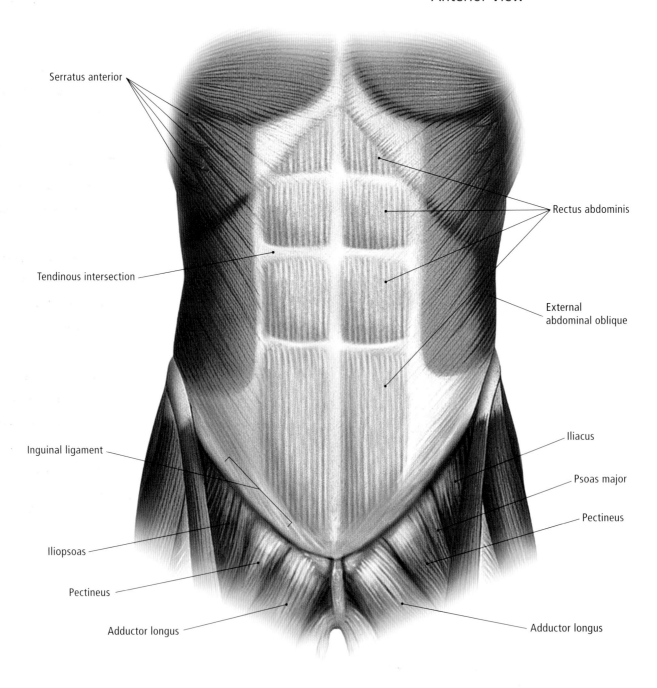

Serratus anterior

Rectus abdominis

Tendinous intersection

External
abdominal oblique

Inguinal ligament

Iliacus

Psoas major

Pectineus

Iliopsoas

Pectineus

Adductor longus

Adductor longus

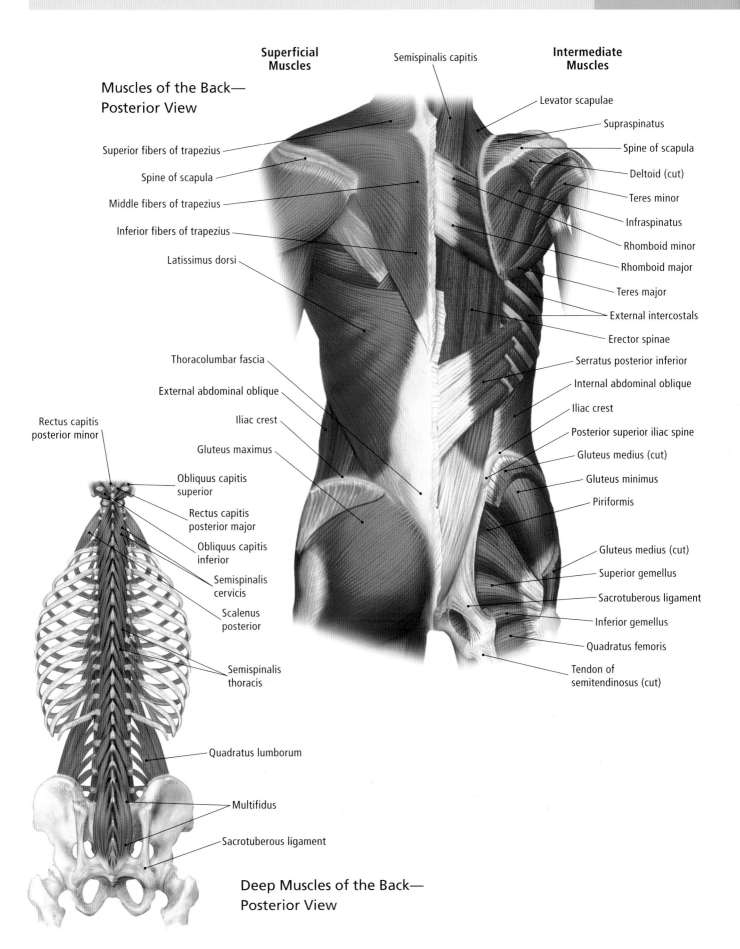

Muscles of the Back—
Posterior View

Superficial Muscles

Semispinalis capitis

Intermediate Muscles

Superior fibers of trapezius

Spine of scapula

Middle fibers of trapezius

Inferior fibers of trapezius

Latissimus dorsi

Thoracolumbar fascia

External abdominal oblique

Iliac crest

Gluteus maximus

Levator scapulae

Supraspinatus

Spine of scapula

Deltoid (cut)

Teres minor

Infraspinatus

Rhomboid minor

Rhomboid major

Teres major

External intercostals

Erector spinae

Serratus posterior inferior

Internal abdominal oblique

Iliac crest

Posterior superior iliac spine

Gluteus medius (cut)

Gluteus minimus

Piriformis

Gluteus medius (cut)

Superior gemellus

Sacrotuberous ligament

Inferior gemellus

Quadratus femoris

Tendon of
semitendinosus (cut)

Rectus capitis
posterior minor

Obliquus capitis
superior

Rectus capitis
posterior major

Obliquus capitis
inferior

Semispinalis
cervicis

Scalenus
posterior

Semispinalis
thoracis

Quadratus lumborum

Multifidus

Sacrotuberous ligament

Deep Muscles of the Back—
Posterior View

Muscles of the Upper and Lower Limb

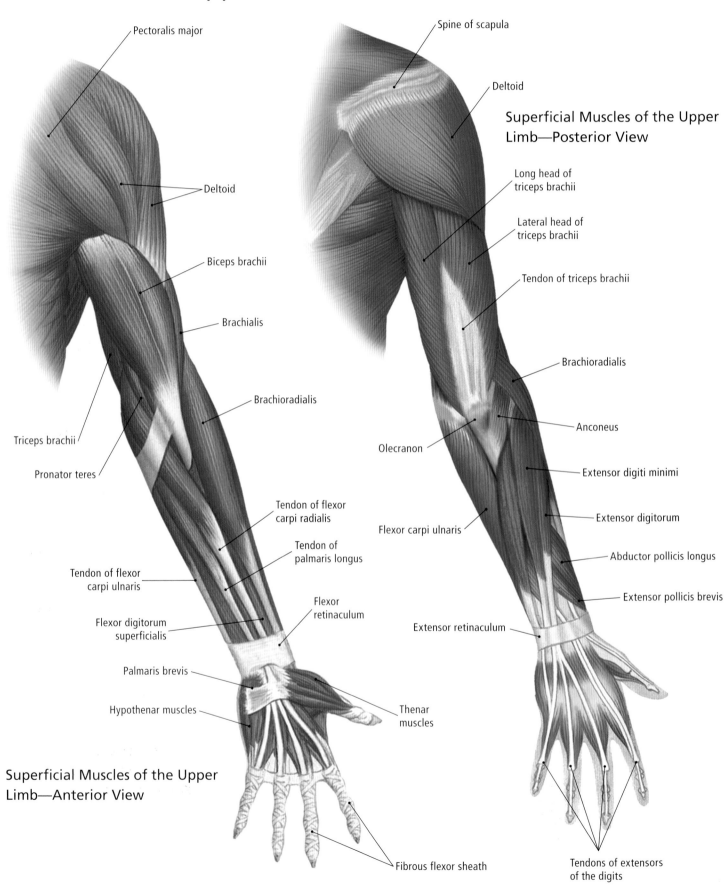

Pectoralis major

Spine of scapula

Deltoid

Superficial Muscles of the Upper Limb—Posterior View

Deltoid

Long head of triceps brachii

Biceps brachii

Lateral head of triceps brachii

Brachialis

Tendon of triceps brachii

Brachioradialis

Brachioradialis

Triceps brachii

Anconeus

Pronator teres

Olecranon

Extensor digiti minimi

Tendon of flexor carpi radialis

Flexor carpi ulnaris

Extensor digitorum

Tendon of palmaris longus

Tendon of flexor carpi ulnaris

Abductor pollicis longus

Flexor digitorum superficialis

Flexor retinaculum

Extensor pollicis brevis

Palmaris brevis

Extensor retinaculum

Hypothenar muscles

Thenar muscles

Superficial Muscles of the Upper Limb—Anterior View

Fibrous flexor sheath

Tendons of extensors of the digits

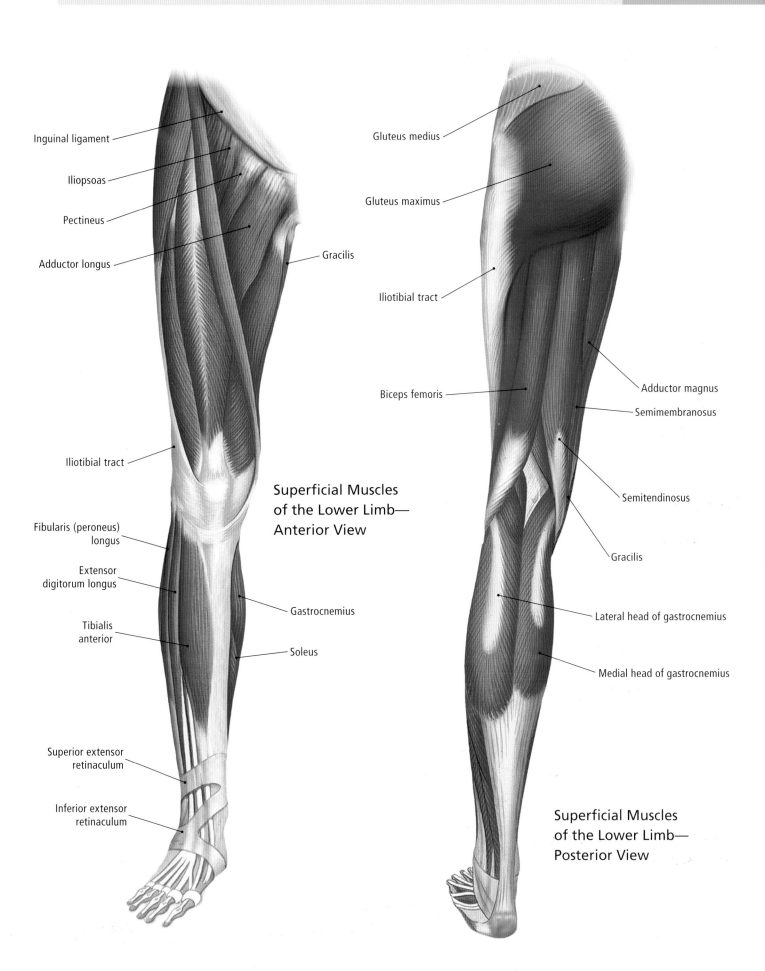

Inguinal ligament

Iliopsoas

Pectineus

Adductor longus

Gracilis

Iliotibial tract

Fibularis (peroneus) longus

Extensor digitorum longus

Tibialis anterior

Gastrocnemius

Soleus

Superior extensor retinaculum

Inferior extensor retinaculum

Superficial Muscles of the Lower Limb— Anterior View

Gluteus medius

Gluteus maximus

Iliotibial tract

Biceps femoris

Adductor magnus

Semimembranosus

Semitendinosus

Gracilis

Lateral head of gastrocnemius

Medial head of gastrocnemius

Superficial Muscles of the Lower Limb— Posterior View

Bones of the Body

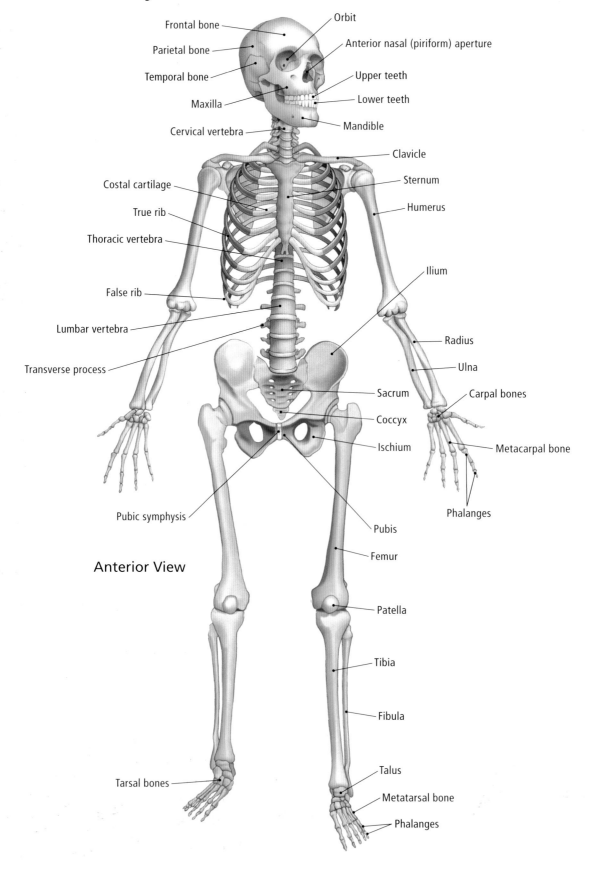

Frontal bone

Orbit

Parietal bone

Anterior nasal (piriform) aperture

Temporal bone

Upper teeth

Maxilla

Lower teeth

Cervical vertebra

Mandible

Costal cartilage

Clavicle

True rib

Sternum

Thoracic vertebra

Humerus

False rib

Ilium

Lumbar vertebra

Transverse process

Radius

Ulna

Sacrum

Carpal bones

Coccyx

Metacarpal bone

Ischium

Pubic symphysis

Phalanges

Pubis

Femur

Anterior View

Patella

Tibia

Fibula

Talus

Tarsal bones

Metatarsal bone

Phalanges

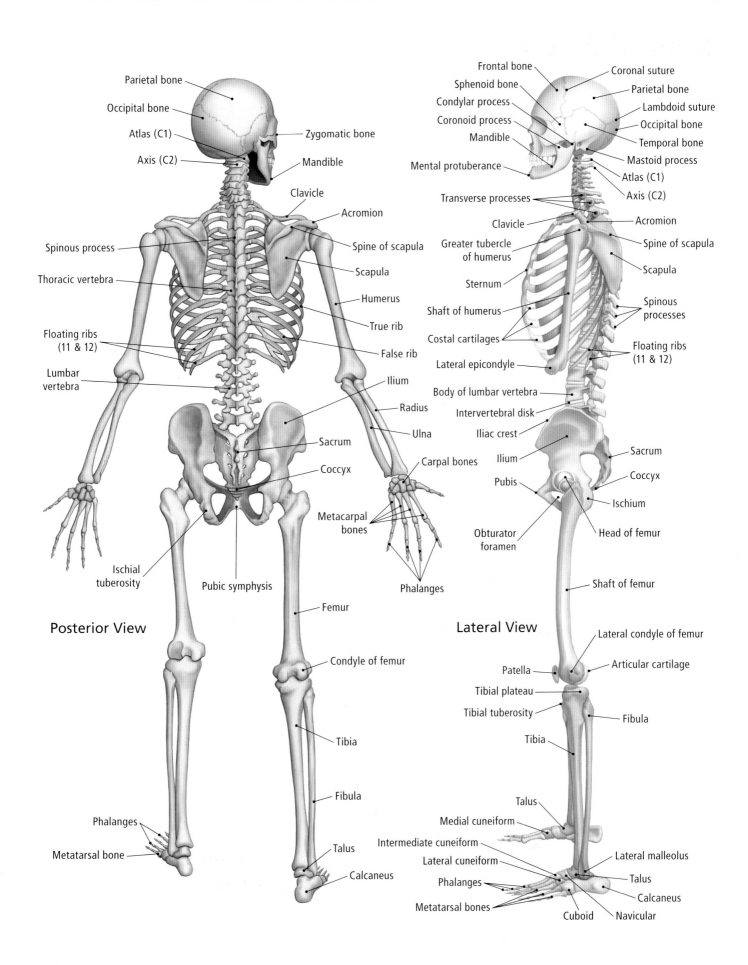

Posterior View

Parietal bone
Occipital bone
Atlas (C1)
Axis (C2)
Zygomatic bone
Mandible
Clavicle
Acromion
Spine of scapula
Spinous process
Scapula
Thoracic vertebra
Humerus
True rib
Floating ribs (11 & 12)
False rib
Lumbar vertebra
Ilium
Radius
Ulna
Sacrum
Coccyx
Carpal bones
Metacarpal bones
Ischial tuberosity
Pubic symphysis
Phalanges
Femur
Condyle of femur
Tibia
Fibula
Phalanges
Talus
Metatarsal bone
Calcaneus

Lateral View

Frontal bone
Sphenoid bone
Condylar process
Coronoid process
Mandible
Mental protuberance
Coronal suture
Parietal bone
Lambdoid suture
Occipital bone
Temporal bone
Mastoid process
Atlas (C1)
Axis (C2)
Transverse processes
Clavicle
Acromion
Spine of scapula
Greater tubercle of humerus
Scapula
Sternum
Spinous processes
Shaft of humerus
Costal cartilages
Floating ribs (11 & 12)
Lateral epicondyle
Body of lumbar vertebra
Intervertebral disk
Iliac crest
Sacrum
Ilium
Coccyx
Pubis
Ischium
Obturator foramen
Head of femur
Shaft of femur
Lateral condyle of femur
Patella
Articular cartilage
Tibial plateau
Tibial tuberosity
Fibula
Tibia
Talus
Medial cuneiform
Intermediate cuneiform
Lateral cuneiform
Lateral malleolus
Phalanges
Talus
Metatarsal bones
Calcaneus
Cuboid
Navicular

Vertebral Column

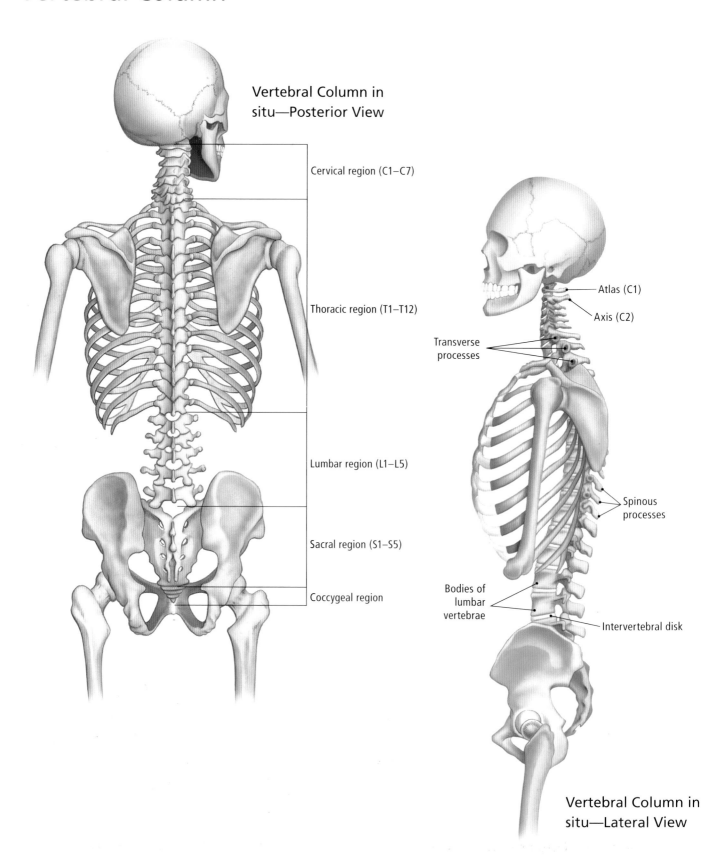

Vertebral Column in situ—Posterior View

Cervical region (C1–C7)

Thoracic region (T1–T12)

Lumbar region (L1–L5)

Sacral region (S1–S5)

Coccygeal region

Atlas (C1)

Axis (C2)

Transverse processes

Spinous processes

Bodies of lumbar vertebrae

Intervertebral disk

Vertebral Column in situ—Lateral View

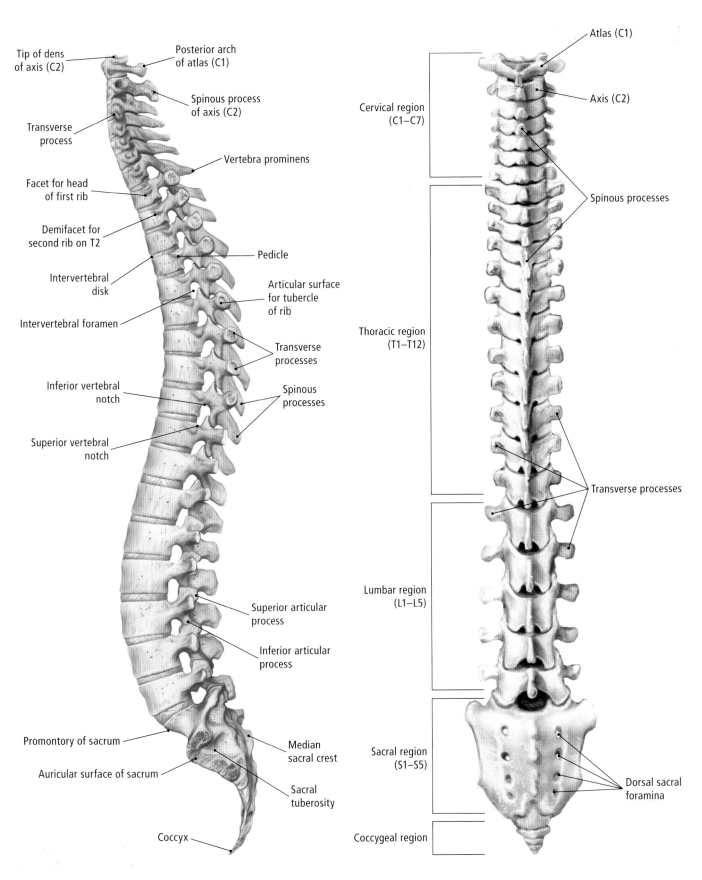

Vertebral Column—Lateral View

Tip of dens of axis (C2)

Posterior arch of atlas (C1)

Spinous process of axis (C2)

Transverse process

Vertebra prominens

Facet for head of first rib

Demifacet for second rib on T2

Pedicle

Intervertebral disk

Articular surface for tubercle of rib

Intervertebral foramen

Transverse processes

Inferior vertebral notch

Spinous processes

Superior vertebral notch

Superior articular process

Inferior articular process

Promontory of sacrum

Median sacral crest

Auricular surface of sacrum

Sacral tuberosity

Coccyx

Vertebral Column—Posterior View

Atlas (C1)

Axis (C2)

Cervical region (C1–C7)

Spinous processes

Thoracic region (T1–T12)

Transverse processes

Lumbar region (L1–L5)

Sacral region (S1–S5)

Dorsal sacral foramina

Coccygeal region

Bones of the Upper and Lower Limb

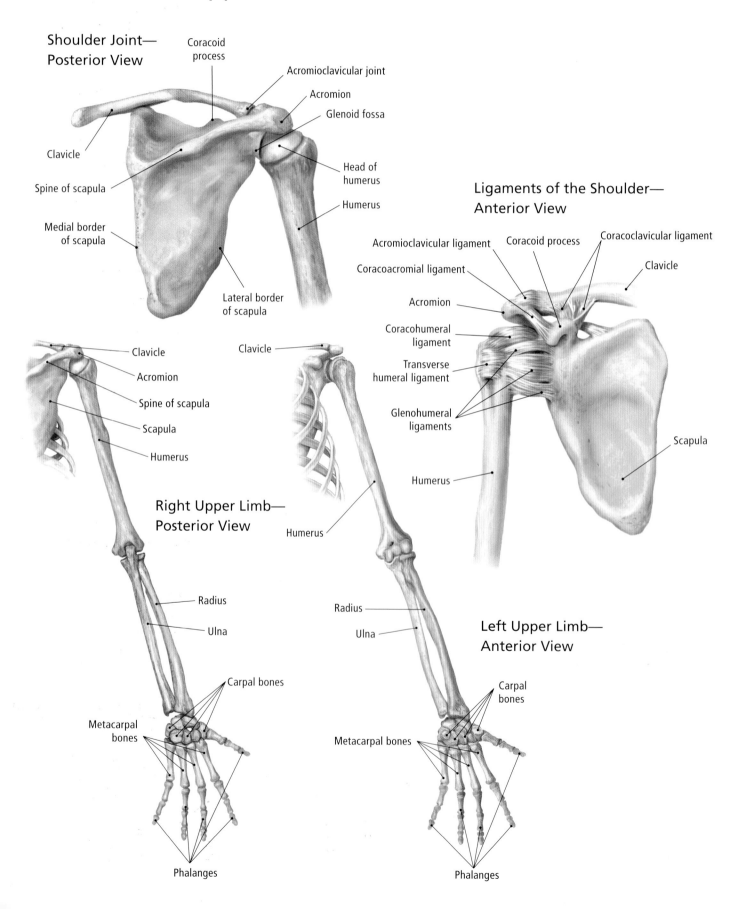

Shoulder Joint—
Posterior View

Coracoid process

Acromioclavicular joint

Acromion

Glenoid fossa

Head of humerus

Humerus

Clavicle

Spine of scapula

Medial border of scapula

Lateral border of scapula

Ligaments of the Shoulder—
Anterior View

Acromioclavicular ligament

Coracoacromial ligament

Acromion

Coracohumeral ligament

Transverse humeral ligament

Glenohumeral ligaments

Coracoid process

Coracoclavicular ligament

Clavicle

Humerus

Scapula

Clavicle

Acromion

Spine of scapula

Scapula

Humerus

Right Upper Limb—
Posterior View

Clavicle

Humerus

Radius

Ulna

Radius

Ulna

Left Upper Limb—
Anterior View

Carpal bones

Metacarpal bones

Carpal bones

Metacarpal bones

Phalanges

Phalanges

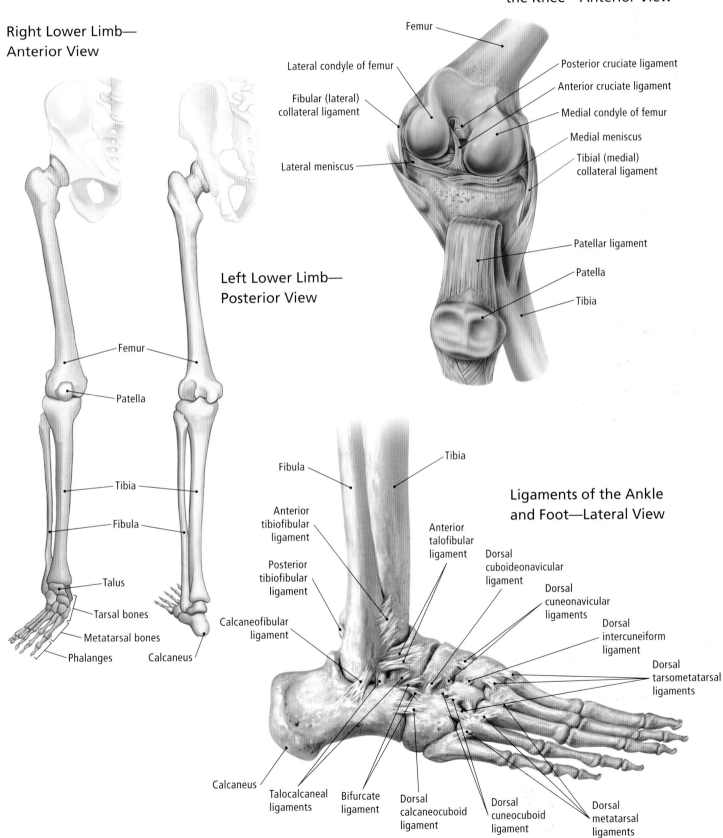

Right Lower Limb—Anterior View

Femur

Patella

Tibia

Fibula

Talus

Tarsal bones

Metatarsal bones

Phalanges

Calcaneus

Left Lower Limb—Posterior View

Bones and Ligaments of the Knee—Anterior View

Femur

Lateral condyle of femur

Fibular (lateral) collateral ligament

Lateral meniscus

Posterior cruciate ligament

Anterior cruciate ligament

Medial condyle of femur

Medial meniscus

Tibial (medial) collateral ligament

Patellar ligament

Patella

Tibia

Ligaments of the Ankle and Foot—Lateral View

Fibula

Tibia

Anterior tibiofibular ligament

Posterior tibiofibular ligament

Calcaneofibular ligament

Anterior talofibular ligament

Dorsal cuboideonavicular ligament

Dorsal cuneonavicular ligaments

Dorsal intercuneiform ligament

Dorsal tarsometatarsal ligaments

Calcaneus

Talocalcaneal ligaments

Bifurcate ligament

Dorsal calcaneocuboid ligament

Dorsal cuneocuboid ligament

Dorsal metatarsal ligaments

Nervous System

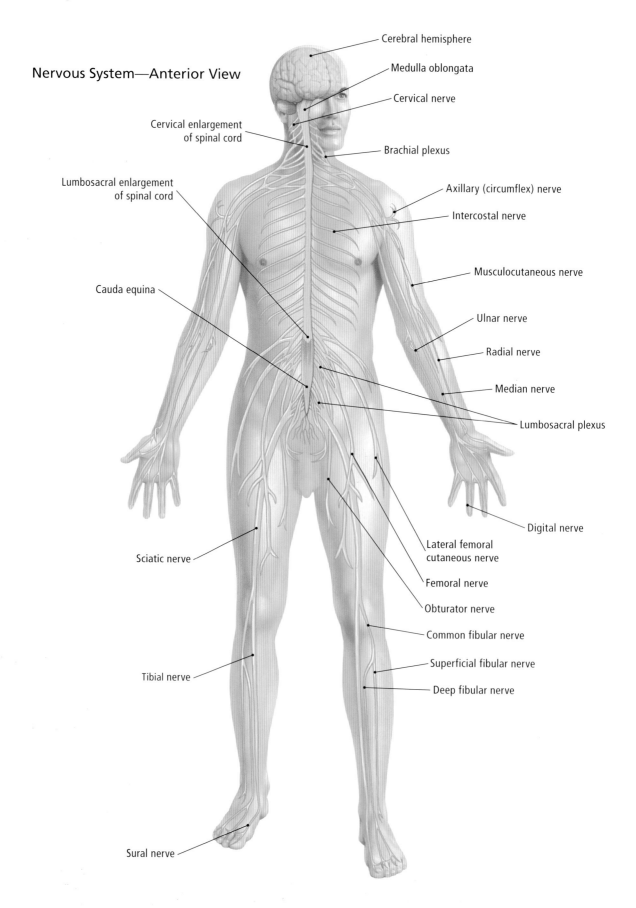

Nervous System—Anterior View

Cerebral hemisphere

Medulla oblongata

Cervical nerve

Cervical enlargement of spinal cord

Brachial plexus

Lumbosacral enlargement of spinal cord

Axillary (circumflex) nerve

Intercostal nerve

Musculocutaneous nerve

Ulnar nerve

Cauda equina

Radial nerve

Median nerve

Lumbosacral plexus

Digital nerve

Sciatic nerve

Lateral femoral cutaneous nerve

Femoral nerve

Obturator nerve

Common fibular nerve

Superficial fibular nerve

Tibial nerve

Deep fibular nerve

Sural nerve

Central Nervous System

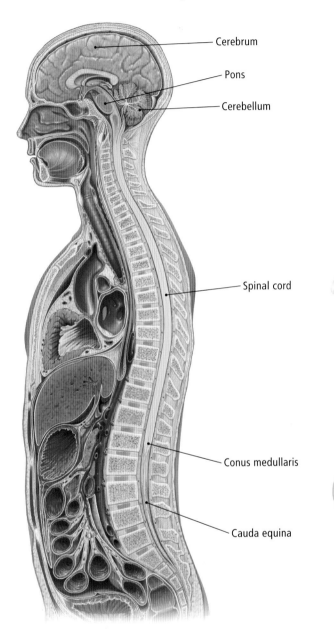

Cerebrum

Pons

Cerebellum

Spinal cord

Conus medullaris

Cauda equina

Autonomic Nervous System

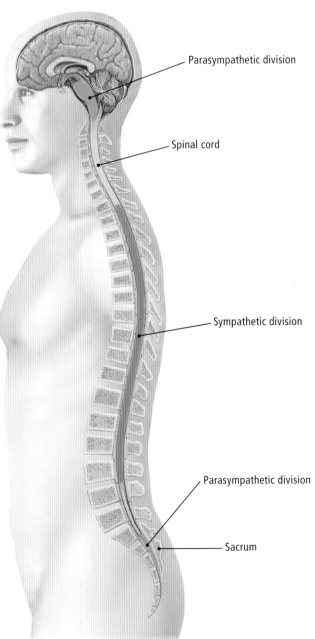

Parasympathetic division

Spinal cord

Sympathetic division

Parasympathetic division

Sacrum

Spinal Cord

Spinal Cord—Cross-sectional View

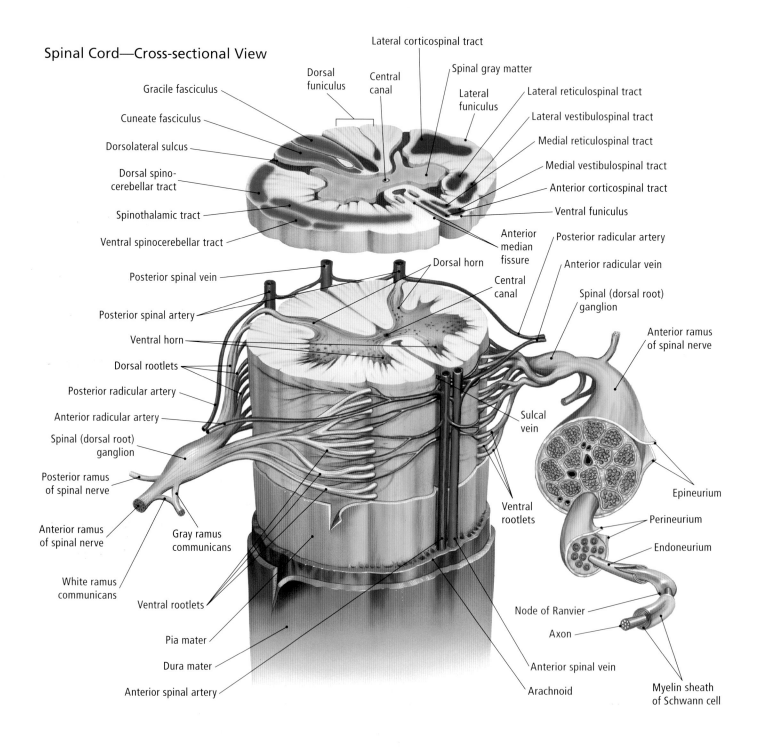

Spinal Nerves

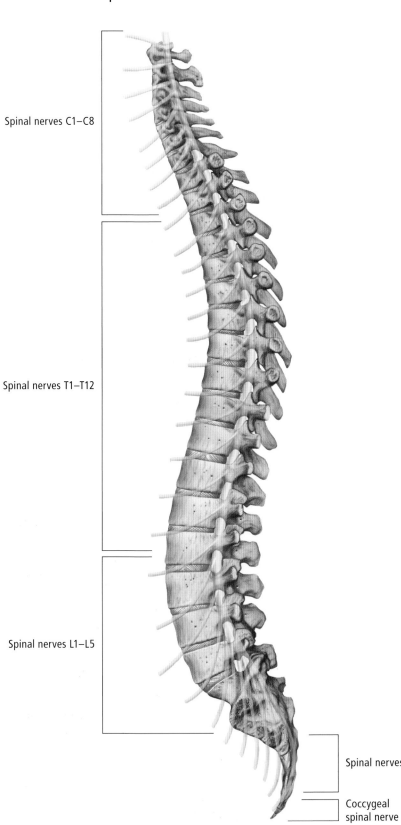

Spinal nerves C1–C8

Spinal nerves T1–T12

Spinal nerves L1–L5

Spinal nerves S1–S5

Coccygeal
spinal nerve

Spinal Cord—Anterior View

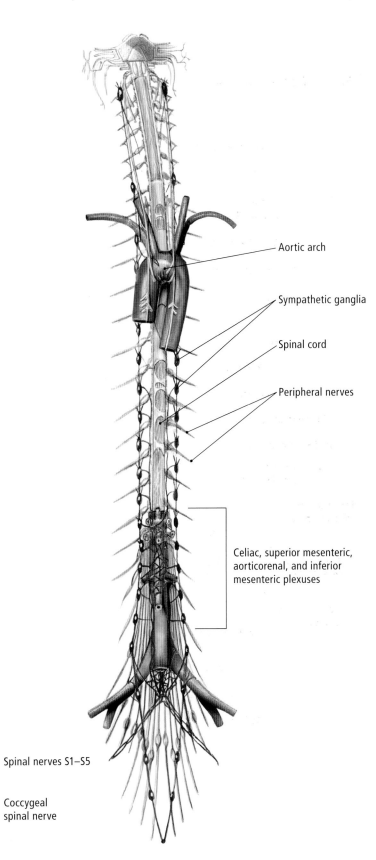

Aortic arch

Sympathetic ganglia

Spinal cord

Peripheral nerves

Celiac, superior mesenteric,
aorticorenal, and inferior
mesenteric plexuses

Circulatory System

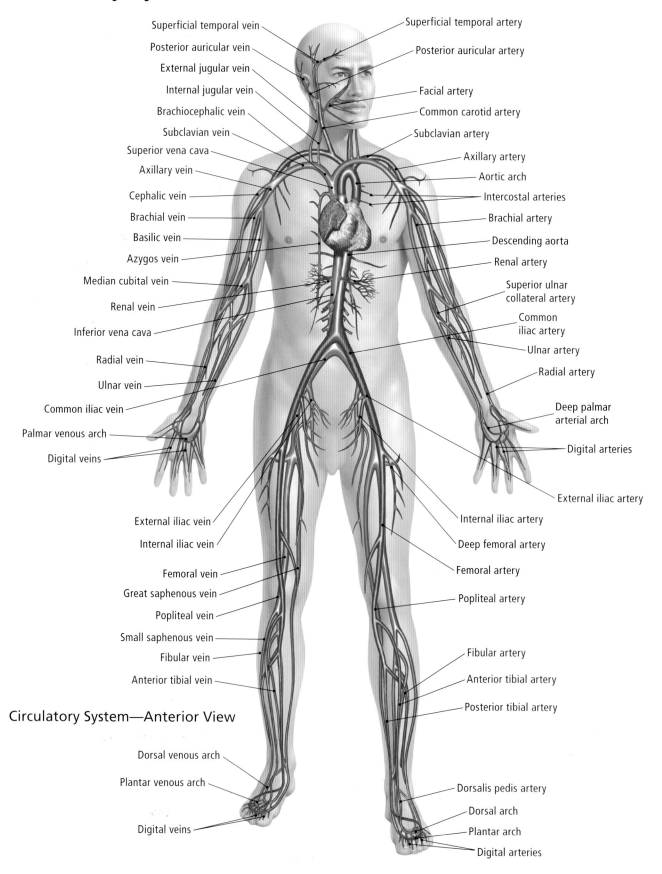

Superficial temporal vein
Posterior auricular vein
External jugular vein
Internal jugular vein
Brachiocephalic vein
Subclavian vein
Superior vena cava
Axillary vein
Cephalic vein
Brachial vein
Basilic vein
Azygos vein
Median cubital vein
Renal vein
Inferior vena cava
Radial vein
Ulnar vein
Common iliac vein
Palmar venous arch
Digital veins

Superficial temporal artery
Posterior auricular artery
Facial artery
Common carotid artery
Subclavian artery
Axillary artery
Aortic arch
Intercostal arteries
Brachial artery
Descending aorta
Renal artery
Superior ulnar collateral artery
Common iliac artery
Ulnar artery
Radial artery
Deep palmar arterial arch
Digital arteries
External iliac artery
Internal iliac artery
Deep femoral artery
Femoral artery
Popliteal artery
Fibular artery
Anterior tibial artery
Posterior tibial artery

External iliac vein
Internal iliac vein
Femoral vein
Great saphenous vein
Popliteal vein
Small saphenous vein
Fibular vein
Anterior tibial vein

Circulatory System—Anterior View

Dorsal venous arch
Plantar venous arch
Digital veins

Dorsalis pedis artery
Dorsal arch
Plantar arch
Digital arteries

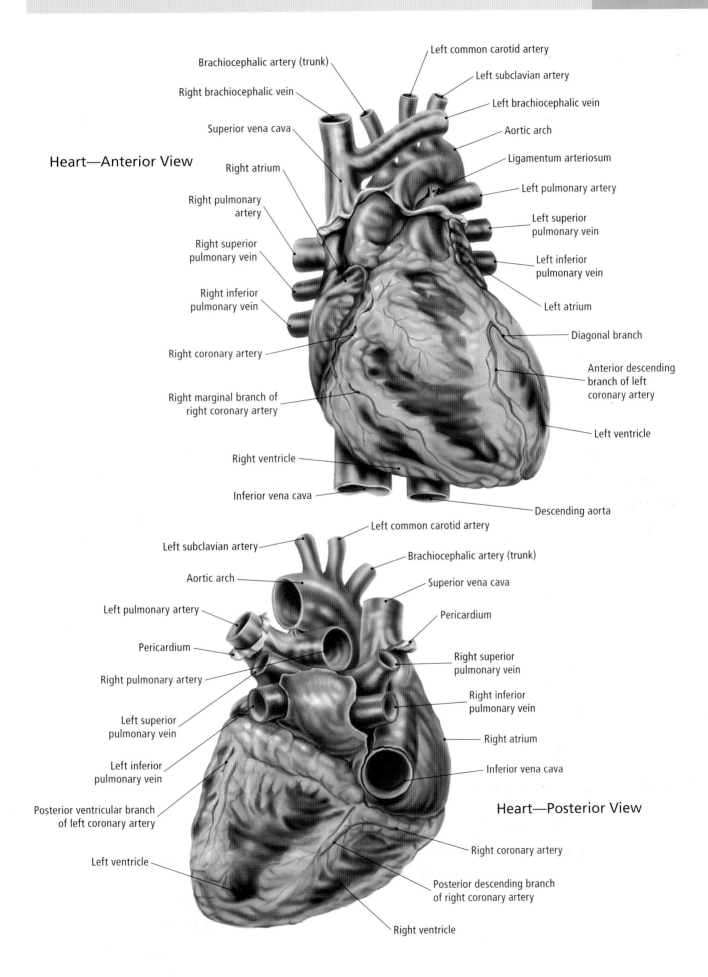

Heart—Anterior View

Brachiocephalic artery (trunk)
Right brachiocephalic vein
Superior vena cava
Right atrium
Right pulmonary artery
Right superior pulmonary vein
Right inferior pulmonary vein
Right coronary artery
Right marginal branch of right coronary artery
Right ventricle
Inferior vena cava

Left common carotid artery
Left subclavian artery
Left brachiocephalic vein
Aortic arch
Ligamentum arteriosum
Left pulmonary artery
Left superior pulmonary vein
Left inferior pulmonary vein
Left atrium
Diagonal branch
Anterior descending branch of left coronary artery
Left ventricle
Descending aorta

Left common carotid artery
Left subclavian artery
Aortic arch
Left pulmonary artery
Pericardium
Right pulmonary artery
Left superior pulmonary vein
Left inferior pulmonary vein
Posterior ventricular branch of left coronary artery
Left ventricle

Brachiocephalic artery (trunk)
Superior vena cava
Pericardium
Right superior pulmonary vein
Right inferior pulmonary vein
Right atrium
Inferior vena cava

Heart—Posterior View

Right coronary artery
Posterior descending branch of right coronary artery
Right ventricle

Blood Vessels of the Upper and Lower Limb

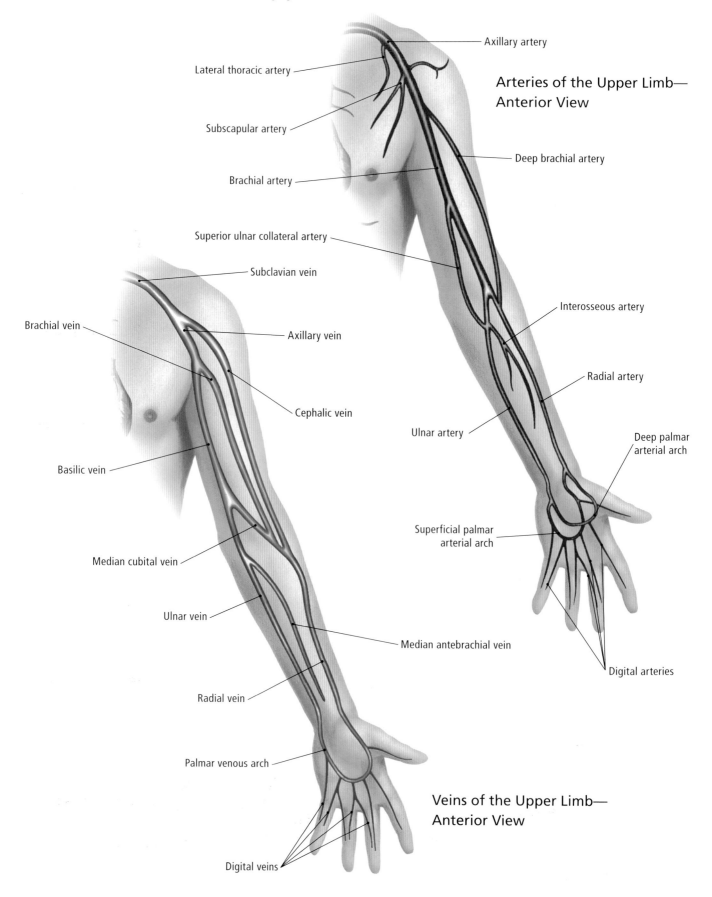

Axillary artery

Lateral thoracic artery

**Arteries of the Upper Limb—
Anterior View**

Subscapular artery

Deep brachial artery

Brachial artery

Superior ulnar collateral artery

Subclavian vein

Interosseous artery

Brachial vein

Axillary vein

Radial artery

Cephalic vein

Ulnar artery

Deep palmar
arterial arch

Basilic vein

Superficial palmar
arterial arch

Median cubital vein

Ulnar vein

Median antebrachial vein

Radial vein

Palmar venous arch

Digital arteries

**Veins of the Upper Limb—
Anterior View**

Digital veins

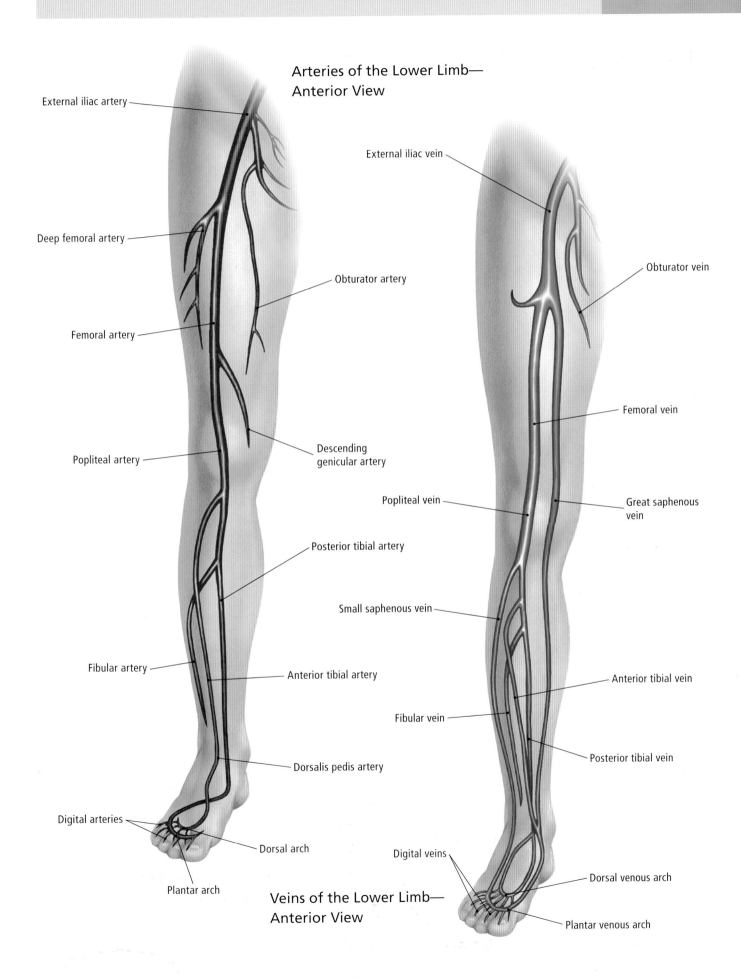

Arteries of the Lower Limb—
Anterior View

External iliac artery

Deep femoral artery

Obturator artery

Femoral artery

Popliteal artery

Descending
genicular artery

Posterior tibial artery

Fibular artery

Anterior tibial artery

Dorsalis pedis artery

Digital arteries

Dorsal arch

Plantar arch

External iliac vein

Obturator vein

Femoral vein

Popliteal vein

Great saphenous
vein

Small saphenous vein

Anterior tibial vein

Fibular vein

Posterior tibial vein

Digital veins

Dorsal venous arch

Plantar venous arch

Veins of the Lower Limb—
Anterior View

Respiratory System

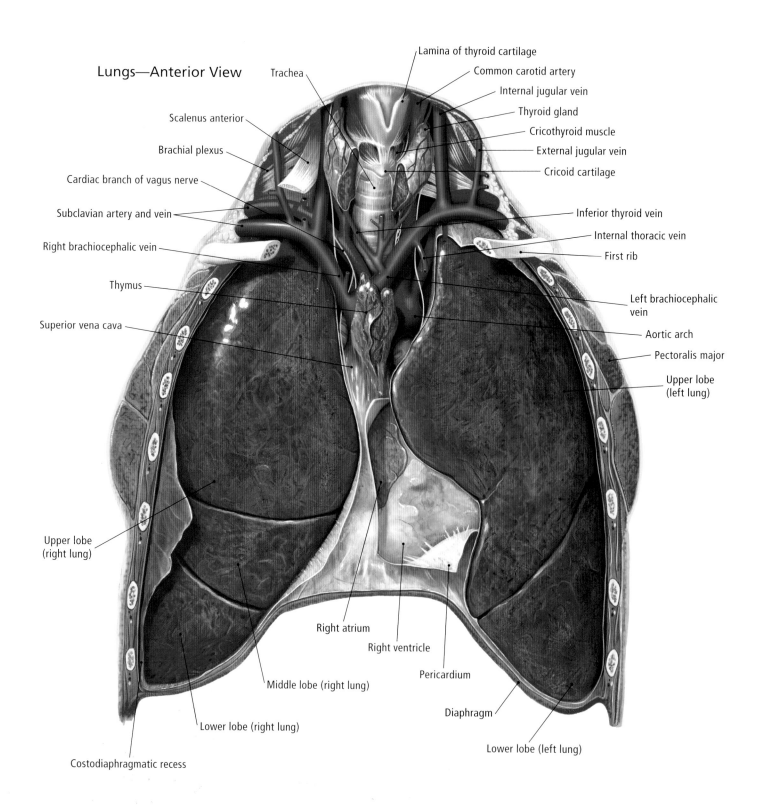

Lungs—Anterior View

- Lamina of thyroid cartilage
- Common carotid artery
- Internal jugular vein
- Thyroid gland
- Cricothyroid muscle
- External jugular vein
- Cricoid cartilage
- Trachea
- Scalenus anterior
- Brachial plexus
- Cardiac branch of vagus nerve
- Subclavian artery and vein
- Right brachiocephalic vein
- Thymus
- Superior vena cava
- Inferior thyroid vein
- Internal thoracic vein
- First rib
- Left brachiocephalic vein
- Aortic arch
- Pectoralis major
- Upper lobe (left lung)
- Upper lobe (right lung)
- Right atrium
- Right ventricle
- Pericardium
- Diaphragm
- Middle lobe (right lung)
- Lower lobe (right lung)
- Lower lobe (left lung)
- Costodiaphragmatic recess

Respiratory System—Anterior View

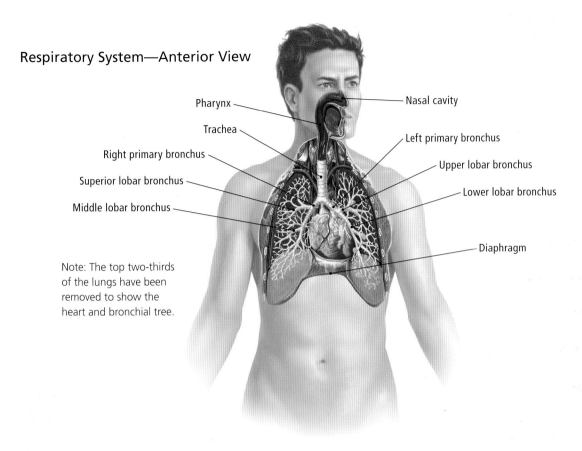

Pharynx — — Nasal cavity

Trachea — — Left primary bronchus

Right primary bronchus — — Upper lobar bronchus

Superior lobar bronchus — — Lower lobar bronchus

Middle lobar bronchus — — Diaphragm

Note: The top two-thirds of the lungs have been removed to show the heart and bronchial tree.

Diaphragm—Inferior View

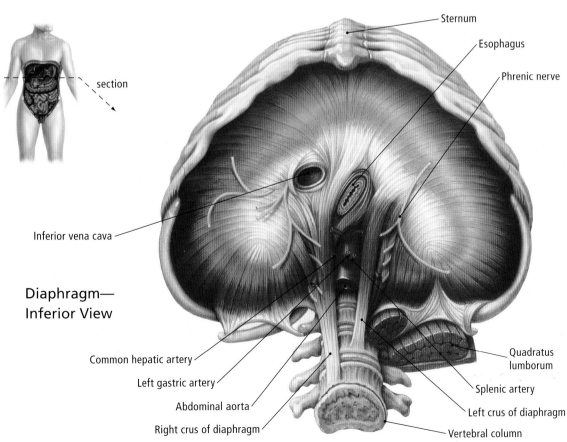

section

Sternum

Esophagus

Phrenic nerve

Inferior vena cava

Common hepatic artery

Left gastric artery

Abdominal aorta

Right crus of diaphragm

Quadratus lumborum

Splenic artery

Left crus of diaphragm

Vertebral column

Movements of the Body

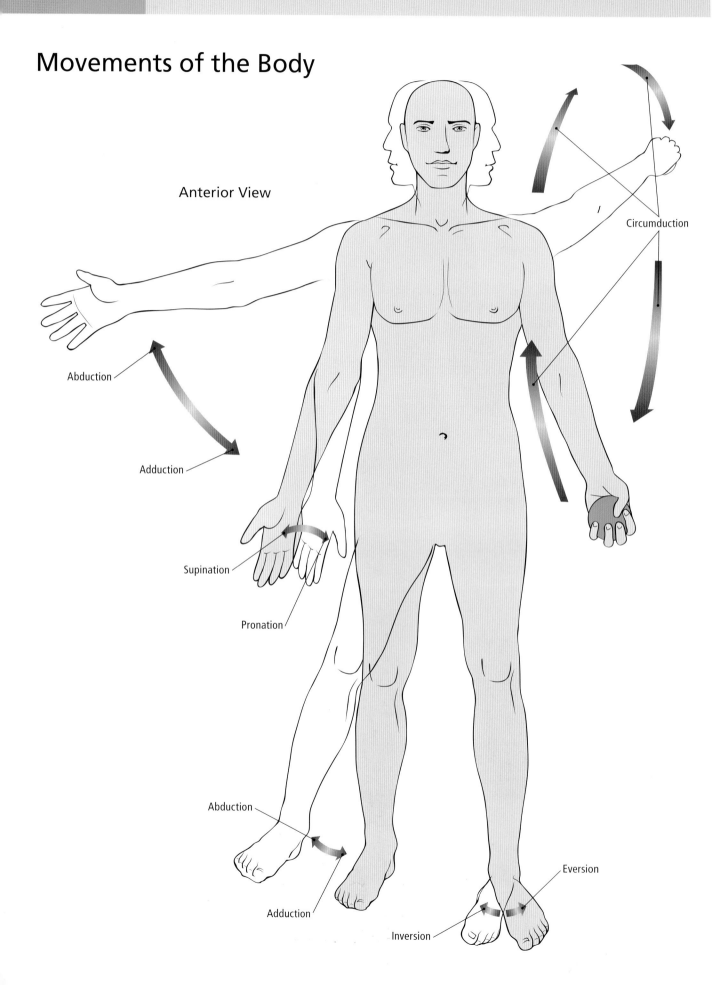

Anterior View

Circumduction

Abduction

Adduction

Supination

Pronation

Abduction

Adduction

Eversion

Inversion

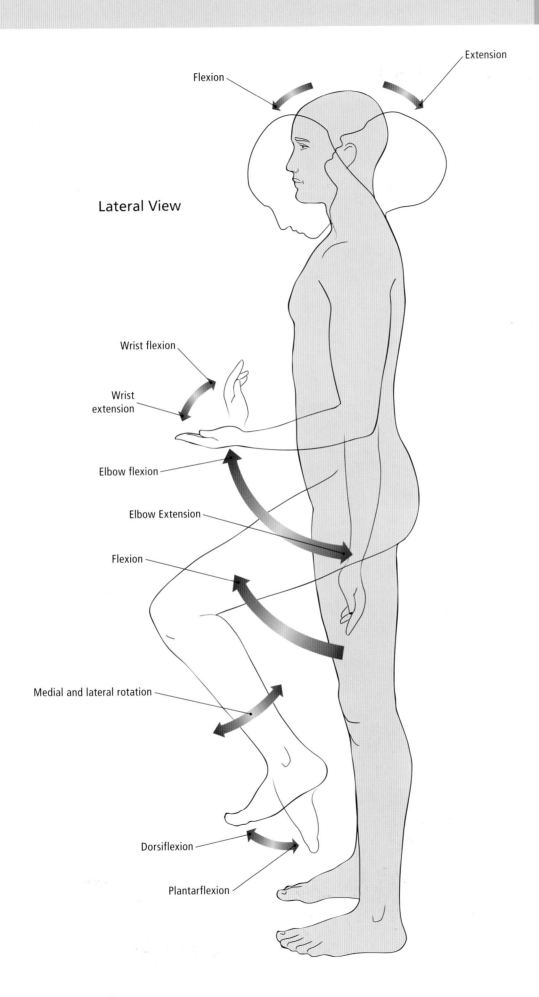

Extension

Flexion

Lateral View

Wrist flexion

Wrist extension

Elbow flexion

Elbow Extension

Flexion

Medial and lateral rotation

Dorsiflexion

Plantarflexion

Exercises

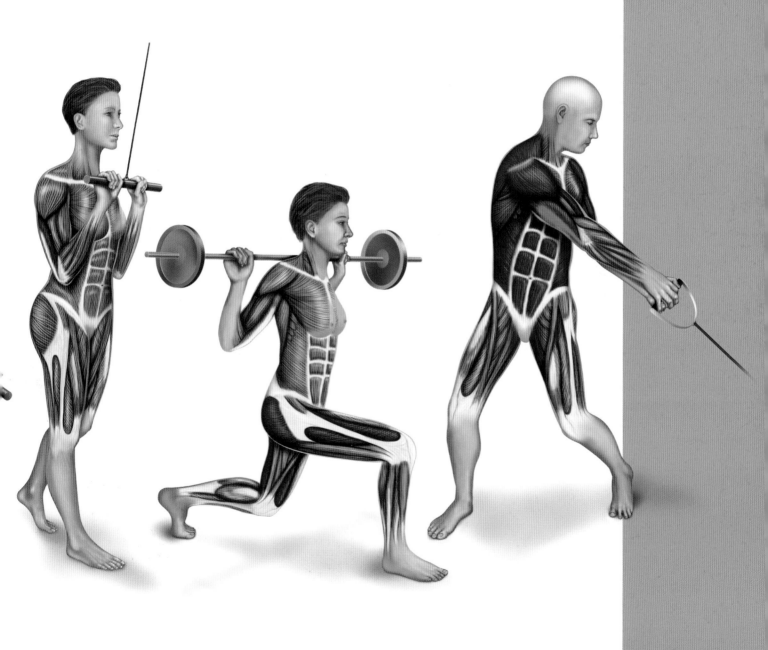

Chest Exercises

A strong chest improves posture, assists breathing, and helps protect the shoulders from injury. The main chest muscle is pectoralis major; other muscles include pectoralis minor, serratus anterior, and the intercostals. Pectoralis major has three primary actions: flexion, adduction, and internal (or medial) rotation of the arm. It is also an accessory muscle of breathing.

Most chest exercises involve pushing, and engage the triceps and deltoids as secondary muscles. Developing powerful pushing muscles makes many tasks easier, meaning less fatigue at the end of the day. For athletes, training chest muscles can help achieve longer and stronger throws, and improve their ability to push off opponents, grapple, and wrestle.

Dumbbell Chest Press

This exercise uses dumbbells to increase the demand on the shoulder and scapular stabilizers. It complements the bench press, as it requires both arms to work independently so that one side of the body cannot "cheat" and let the other side do extra work. The dumbbell chest press targets the pectoral, deltoid, and tricep muscles, with a large stability component required from the rotator cuff, serratus anterior, rhomboids, trapezius, and latissimus dorsi muscles. This exercise helps improve performance in many everyday tasks, such as lifting and pushing, and is ideal to include if training for contact sports, throwing sports, or gymnastics.

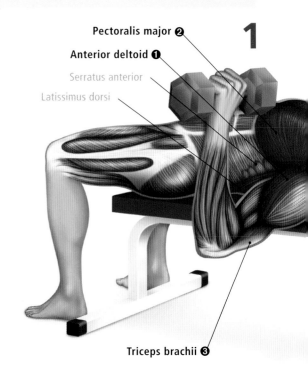

Pectoralis major ❷
Anterior deltoid ❶
Serratus anterior
Latissimus dorsi

1

Triceps brachii ❸

how to

To get into position, sit on the end of a bench with dumbbells resting on the knees. "Kick" the weights up to the shoulders, then lie back flat on the bench. (The "kick" is unnecessary if starting with light weights.) Hold the dumbbells at the sides of the chest, with palms facing down toward the body and elbows bent. Push up and extend the elbows following a small arc of motion so that the dumbbells come close together above the chest. Slowly lower weights until a slight stretch is felt in the front of the shoulder, then repeat.

warning

Do not drop the weights from a lying position when finished, as this can lead to a shoulder dislocation.

variations

EASY
Limit the range of motion in this exercise and do not lower weights past 90 degrees at the elbow if there is a history of shoulder injury. This shorter range of motion decreases the role of the shoulder stabilizers, but still allows heavy weights to be used.

HARD
Use an inclined bench instead of a flat bench and reduce the amount of weight being lifted. This setup still gives the chest muscles a great workout, but the change in bench angle increases the workload of the shoulder and triceps muscles at the same time.

active muscles

❶ Anterior deltoid
❷ Pectoralis major
❸ Triceps brachii

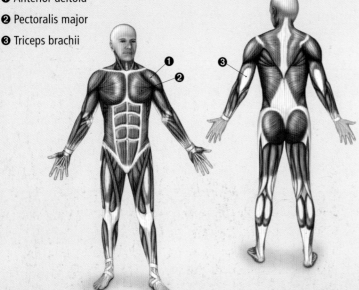

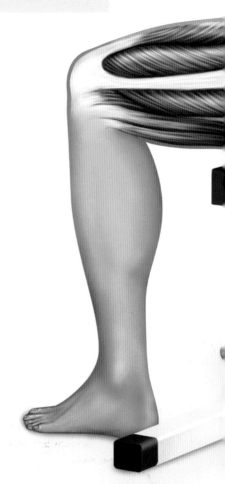

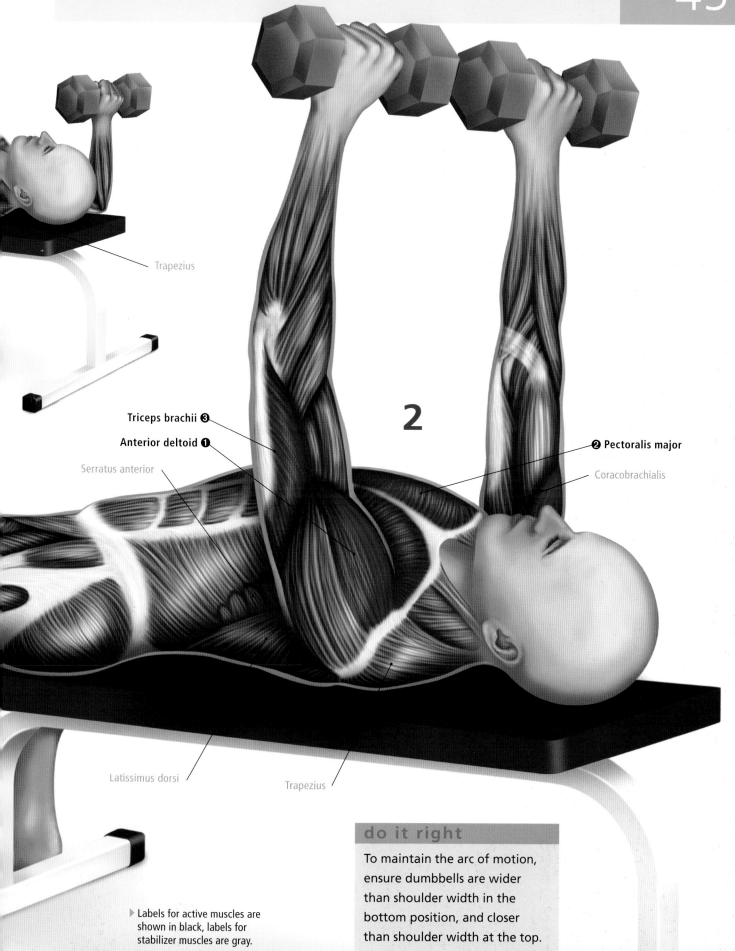

Trapezius

Triceps brachii ❸

Anterior deltoid ❶

Serratus anterior

2

❷ Pectoralis major

Coracobrachialis

Latissimus dorsi

Trapezius

▶ Labels for active muscles are shown in black, labels for stabilizer muscles are gray.

do it right

To maintain the arc of motion, ensure dumbbells are wider than shoulder width in the bottom position, and closer than shoulder width at the top.

Dumbbell Fly

1

This chest-building exercise requires the arm to move through an arc while the elbow remains at a constant angle. The pectoralis major muscle is the primary mover, with the deltoid called on to assist, and the elbow flexors—biceps brachii, brachioradialis, and brachialis—isometrically active. During the eccentric phase of the movement, the chest and arm muscles are also stretched. Because of the long lever arm, the amount of weight lifted to perform the dumbbell fly is significantly less than an equivalent chest exercise, such as the bench press. While this exercise is not as sports specific as the bench press, it is useful as a late-stage rehabilitation exercise for shoulder or elbow injuries.

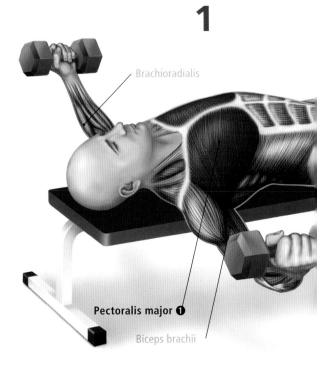

Brachioradialis

Pectoralis major ❶

Biceps brachii

how to

Lie back on a bench and start with hands in a neutral position, gripping the dumbbells above the middle of the chest. The elbows should remain slightly bent. Lower the dumbbells out to the sides of the chest, keeping the elbow angle constant, until a stretch is felt across the front of the shoulders and chest. Bring dumbbells back together at the mid-point above the chest.

variations

EASY

Lie on the floor instead of on a bench. This limits the range of motion and decreases the stretch on the shoulders. Muscles are weakest at their longest and shortest positions, so performing this exercise in the floor position also prevents the pectoralis major from stretching to the end of its range.

HARD

Try one-arm alternating dumbbell flys to isolate each side of the chest. Follow the same movement as for the standard dumbbell fly, but use one arm first, then alternate to the other. This variation provides an additional stability challenge, which can be increased again by performing the exercise lying on a stability ball, rather than on a bench.

active muscles

❶ Pectoralis major

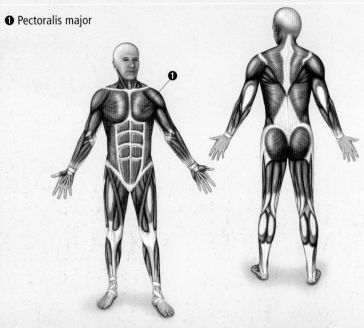

❶

warning

The end position is stressful and unstable for the shoulder joint. Do not risk injury by lifting weights that are too heavy.

do it right

Do not lock elbows. Maintain some elbow flexion throughout the movement but keep the elbow angle constant.

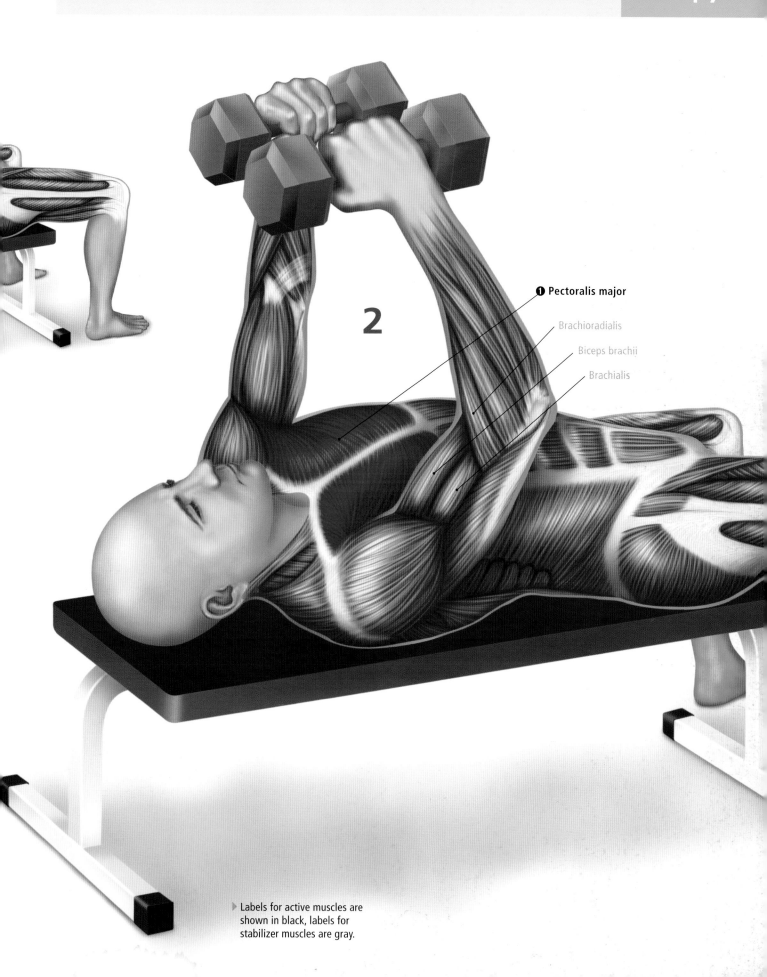

❶ **Pectoralis major**

Brachioradialis

Biceps brachii

Brachialis

2

▶ Labels for active muscles are
shown in black, labels for
stabilizer muscles are gray.

Bench Press

This classic compound exercise involves multiple joints and targets the chest, shoulder, and arm muscles. Heavy bench presses build size and strength in the pectoralis major, anterior deltoid, and triceps brachii, while also developing the rotator cuff muscles and scapular stabilizers. This exercise can be performed by both advanced and beginner weight trainers, because the basic movement is straightforward. It forms the basis of just about every upper-body strengthening program, sports-specific conditioning program, and advanced-rehabilitation program.

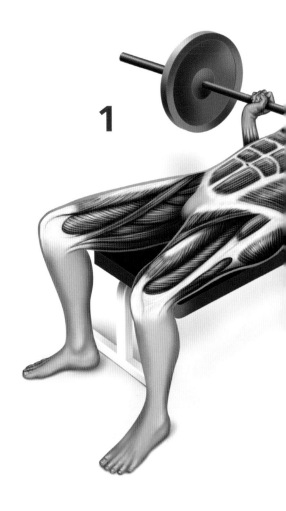

1

how to

Lie back on a flat bench and grasp the bar with both hands a little wider than shoulder-width apart, at an even distance from the weight at each end. Palms face up toward the ceiling. Take a deep breath in, then, while exhaling, lift the bar from the rack so that arms are fully extended and elbows locked. On the next breath in, slowly lower the bar toward the chest. Stop just short of touching the chest. Breathe out and push the weight straight up, away from the chest, returning to full elbow extension.

variations

EASY

Use just the barbell, without any additional weight, if learning the bench press. Focus on technique, and master smooth, controlled repetitions before loading up the bar.

HARD

Try a "close grip" bench press. First, significantly reduce the weight from that of a normal bench press, as the close grip decreases the role of pectoralis major and the deltoid, to focus the work on your triceps muscle. Grasp the bar slightly closer than shoulder-width apart. Lower the weight toward the chest, keeping elbows close to the body. Ensure the grip is not too close, as this will destabilize the bar.

active muscles

❶ Pectoralis major

❷ Anterior deltoid

❸ Triceps brachii

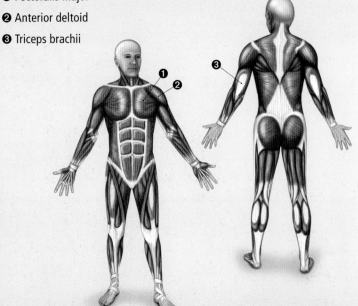

do it right

Do not arch the back with heavy lifts. This is a sign that the weight is too heavy and there is risk of injury.

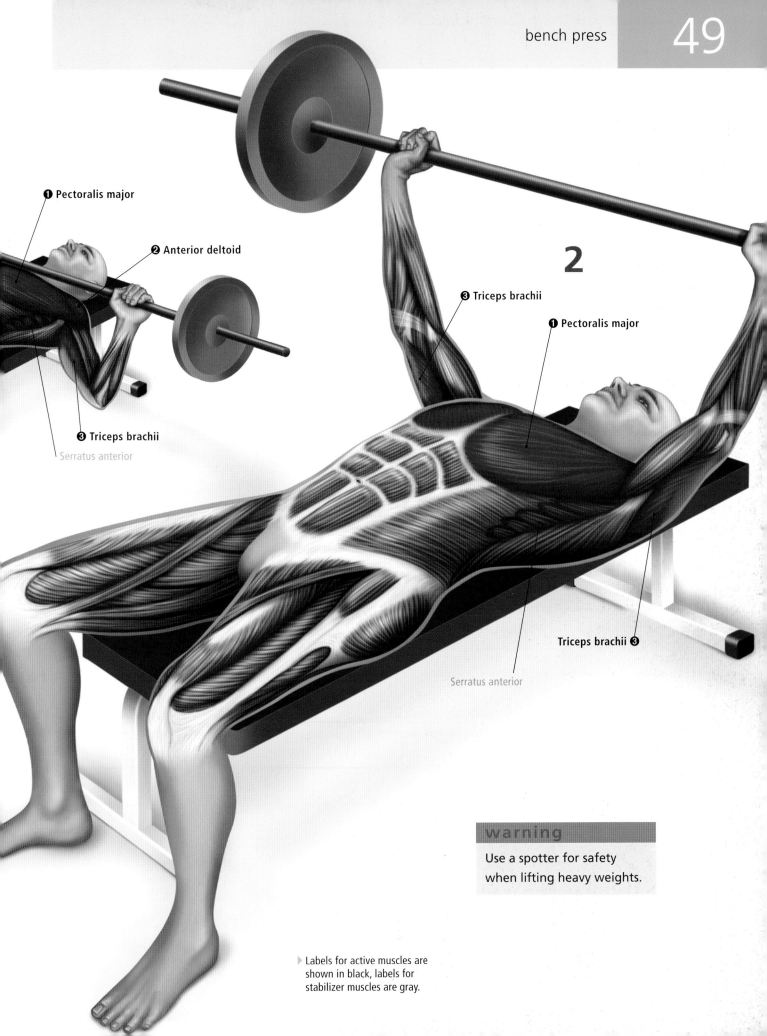

❶ Pectoralis major

❷ Anterior deltoid

❸ Triceps brachii

Serratus anterior

2

❸ Triceps brachii

❶ Pectoralis major

Triceps brachii ❸

Serratus anterior

▶ Labels for active muscles are shown in black, labels for stabilizer muscles are gray.

Dip

This bodyweight exercise complements the pull-up or chin-up, as far as upper body bodyweight exercises are concerned. The dip is a pushing exercise and, as such, it primarily targets the pectoralis major, triceps, and anterior deltoids. Alternatively, it may be possible to vary the dip to proportionally focus more on the triceps or pectoralis major, depending on the amount of forward lean in the trunk. Performed on parallel dip bars, this exercise is difficult for beginners because it requires trainers to have enough strength to lift their own bodyweight, and strong abdominal muscles are needed to stabilize the core throughout the movement. Some gyms have an assisted-dip (and chin-up) machine that helps trainers who do not have the necessary strength to perform the dip unassisted.

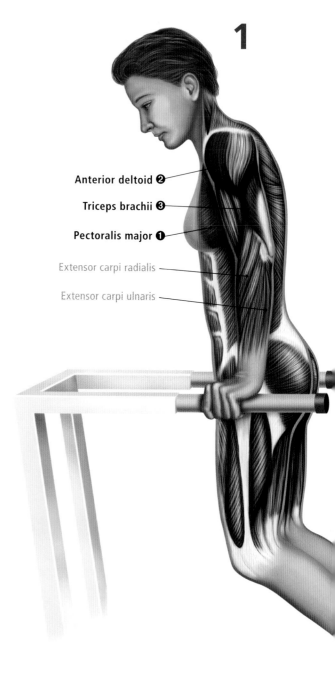

Anterior deltoid ❷

Triceps brachii ❸

Pectoralis major ❶

Extensor carpi radialis

Extensor carpi ulnaris

how to	Stand between the parallel dip bars and use the step so that the bars are at waist height. Grip the bars, lock the elbows, bend the knees, and lift the feet off the ground. Slowly lower the body using eccentric control of the extensor muscles, until elbows are bent to 90 degrees. Push back up until elbows are completely locked again.
variations — EASY	Bench dips are a good alternative if lacking upper body strength or access to an assisted-dip machine. Sit on a bench with legs extended out in front and heels together on the floor. Lift the upper body, supported by the arms. Shuffle forward so that the body hovers in front of the bench, not above it. Slowly lower the body as per the standard dip, then push back up again until elbows are locked.
variations — HARD	Increase the challenge by adding more weight. Use a dip belt, which enables the attachment of extra weight that hangs between the legs.

active muscles

❶ Pectoralis major
❷ Anterior deltoid
❸ Triceps brachii

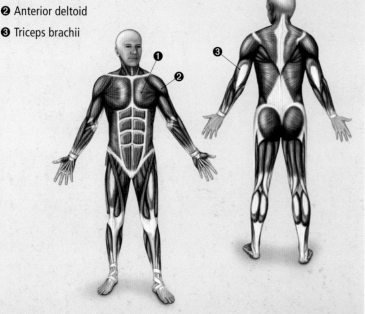

warning

Adding significant extra weight to the dip can cause pectoralis major and triceps tears. Do not add too much weight too quickly.

▶ Labels for active muscles are shown in black, labels for stabilizer muscles are gray.

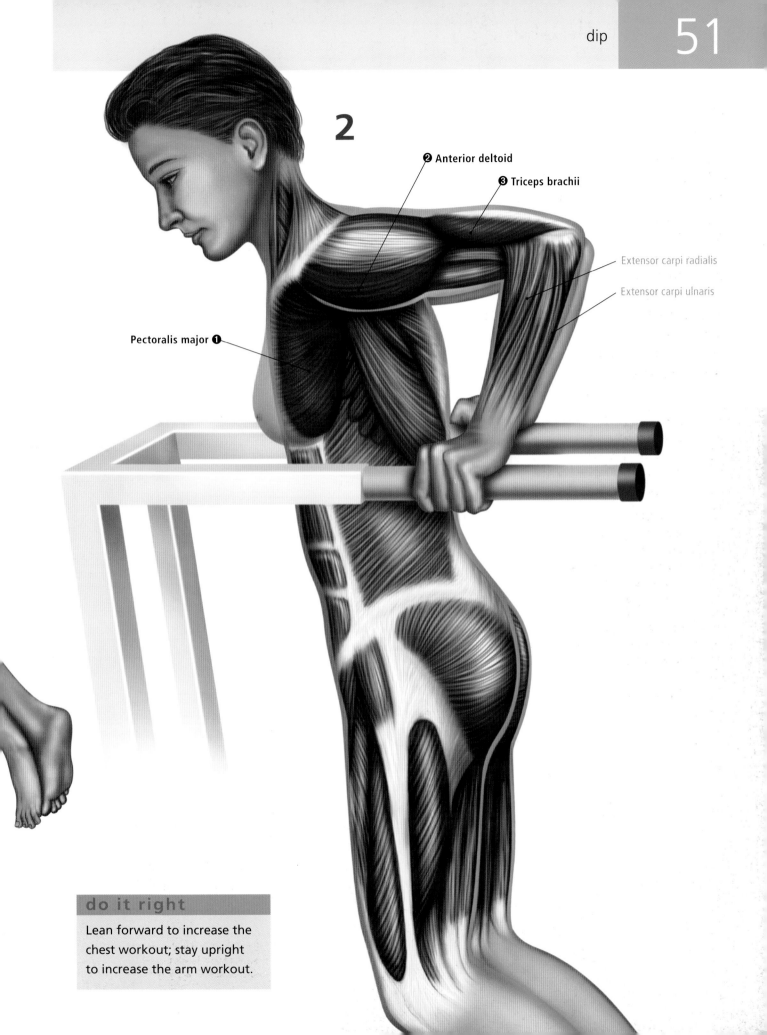

2

❷ Anterior deltoid

❸ Triceps brachii

Extensor carpi radialis

Extensor carpi ulnaris

Pectoralis major ❶

do it right

Lean forward to increase the chest workout; stay upright to increase the arm workout.

Cable Crossover

This exercise, also known as the cable fly, isolates the chest muscles. The concentric phase of the movement helps strengthen the chest, while the eccentric phase provides a good stretch for both the chest and shoulders. Many trainers perform super set cable crossovers with other chest exercises, such as the bench press or dumbbell fly. Use this exercise as an alternative to the dumbbell fly to add variety to routines. The cable crossover is a good strengthening exercise for throwing sports, because of its unilateral nature and its similarity to the throwing motion. Adjust cable height and body position to feel different parts of the muscle working, or to make the movement more sport specific.

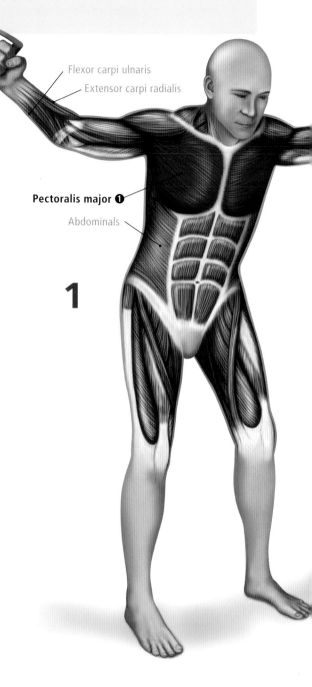

Flexor carpi ulnaris
Extensor carpi radialis
Pectoralis major ❶
Abdominals

1

how to

Stand in the middle of a cable-crossover machine with the pulleys adjusted to above-head height, equal on both sides. Grasp handles with palms facing down, shoulders internally rotated, and hips bent slightly forward. Squeeze the chest muscles to pull arms downward and inward in a hugging motion. Keep elbows at a constant angle throughout the movement. Return to the start position with a slow, controlled motion.

variations

EASY

Lower pulleys to just below shoulder height and perform the standard exercise as described above. If the weight is too heavy, then technique will suffer and the effectiveness of the exercise will be lost. Use light to moderate weights and focus on form at all times.

HARD

Mid- and high-cable crossovers provide an all-around chest workout for advanced weight trainers. For mid crossovers, stand up straight and horizontally flex the arms to meet together in front of the chest. For high crossovers, start below the shoulders and bring hands together above the head.

active muscles

❶ Pectoralis major

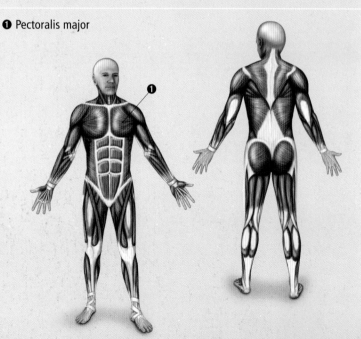

warning

Do not let the arms be pulled back too quickly, as this can cause shoulder dislocations.

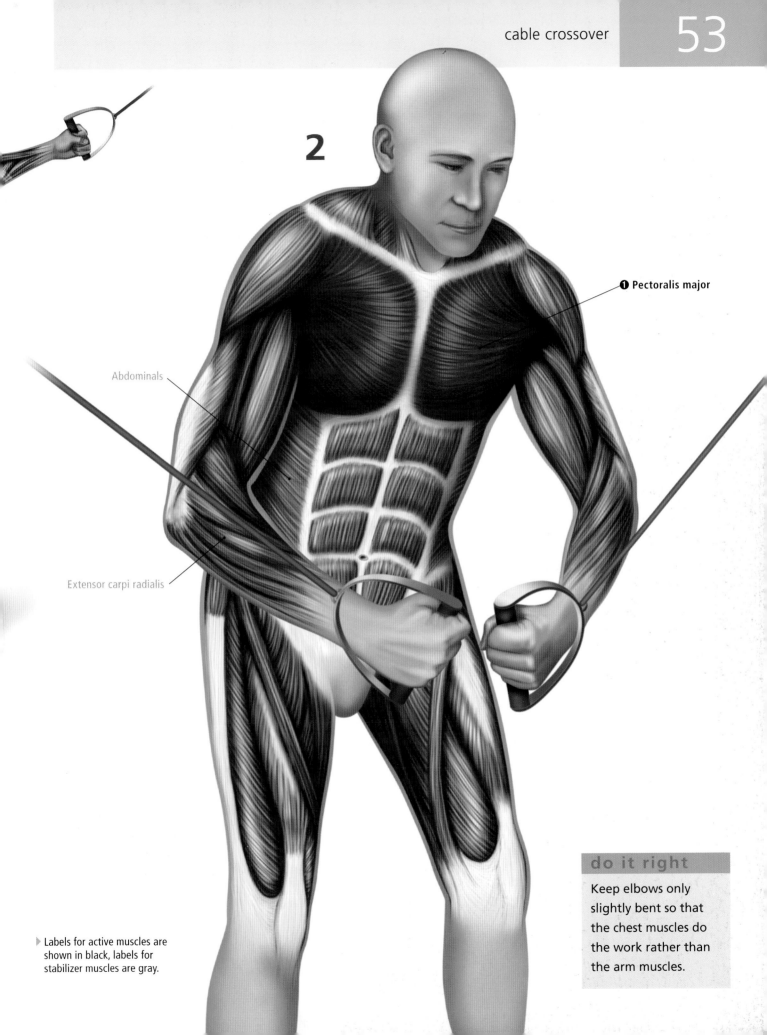

2

❶ **Pectoralis major**

Abdominals

Extensor carpi radialis

▶ Labels for active muscles are shown in black, labels for stabilizer muscles are gray.

do it right
Keep elbows only slightly bent so that the chest muscles do the work rather than the arm muscles.

Pullover

This chest and back exercise creates a great stretch for the pectoral muscles, the lateral muscles, and the abdominal muscles. Beginners may initially find the pullover difficult and intimidating, as the movement requires trainers to lift a dumbbell over the face. For this reason, it is best to start off with a light weight. Slowly build up the weight once completely comfortable with the technique. The pullover is an excellent exercise for throwing sports, as it develops strength and power. The stretching component will improve posture in anyone who has tight shoulders or an increased thoracic kyphosis—office workers typically fit this description.

warning

Range of motion varies according to individual flexibility; do not overstretch.

do it right

Keep elbows slightly bent and in line with the shoulders throughout the movement.

how to

Sit on a bench and grasp a dumbbell at one end with hands together, palms facing up. Lie back carefully, keeping the lower back in a neutral position. Push arms out until fully extended, holding the weight directly above the face. Slowly take the dumbbell back over the head. Keep abdominals tight to maintain a neutral spine. Reach back until a stretch is felt in the chest and shoulders. Keeping the elbows straight, pull the weight up to the starting position.

variations

EASY
Many gyms have a pullover machine that allows the same movement to be performed as the standard exercise, but without the need for core control, or the injury risk of dropping the dumbbell.

HARD
Try a standing cable pulldown. This variation increases the demand on core stability and works the posterior shoulder and the latissimus dorsi muscles. Maintaining a neutral spine while performing standing pulldowns is a significant challenge.

active muscles

❶ Pectoralis major

❷ Latissimus dorsi

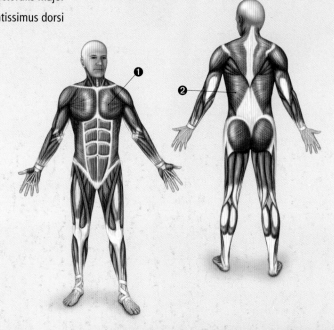

2

▶ Labels for active muscles are shown in black, labels for stabilizer muscles are gray.

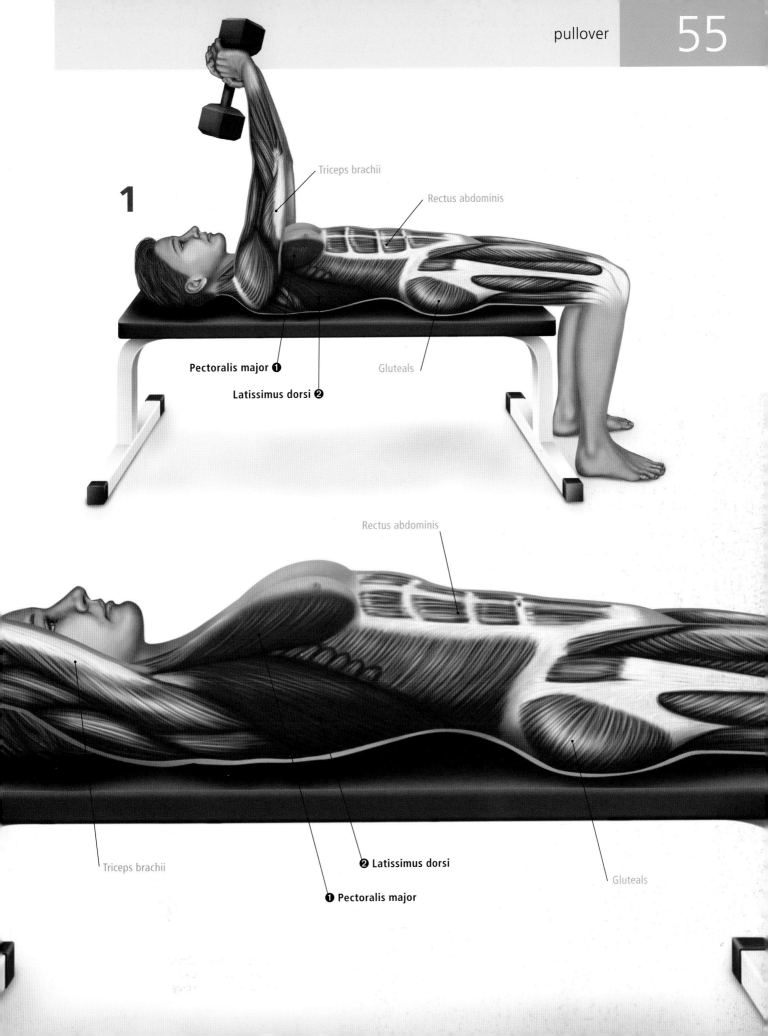

1

Triceps brachii

Rectus abdominis

Pectoralis major ❶

Latissimus dorsi ❷

Gluteals

Rectus abdominis

Triceps brachii

❷ Latissimus dorsi

❶ Pectoralis major

Gluteals

Push-up

This classic exercise is highly effective for improving the strength of the entire body. While the push-up primarily targets muscles in the chest, arms, and shoulders, it also requires support from other muscles. Because a wide range of muscles are integrated into the exercise, the push-up builds both upper body and core strength. It benefits the abdominal muscles by simultaneously flexing and stretching them. When the lower back muscles contract to stabilize the form, the abdominal muscles are inadvertently stretched. The quadriceps are also relied on to maintain proper form, giving the legs a secondary workout. Include the push-up in routines to stabilize the shoulders, as it develops both the scapular and rotator cuff muscles. This exercise does not require any equipment, so the push-up is well suited to daily maintenance routines.

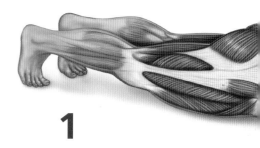

1

how to

To start, lie facedown on the ground with hands beside shoulders, fingers parallel to the body, and feet on their toes. Straighten arms, lifting the body and legs off the ground. Return to the starting position by bending arms and gently lowering the body until it hovers just above the ground.

variations

EASY

Place knees on the floor in the starting position if lacking a high level of upper body strength. Create a plane from the head to the knees as the push-up is performed. Ensure the body does not bend at the hips, as this causes the exercise to lose its effectiveness.

HARD

Place hands together under the body to focus on the triceps, or place them further away from the shoulders to target the chest muscles. While performing a series of basic push-ups, raise each leg in turn to work the lower back and gluteal muscles.

active muscles

❶ Anterior deltoid
❷ Pectoralis major
❸ Serratus anterior
❹ Triceps brachii

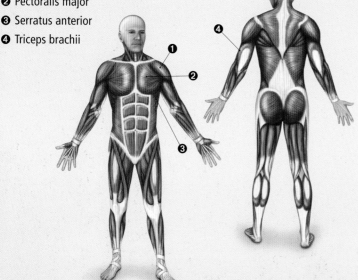

warning

Lifting shoulders in a shrug position as the push-up is performed can destabilize the arms.

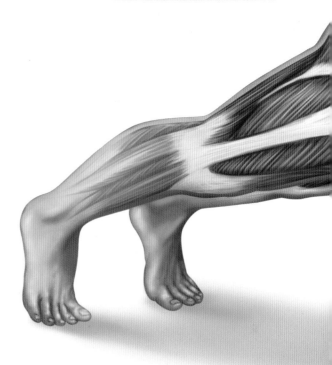

do it right

When lifting into the push-up position, keep body in a flat plane from head to ankles.

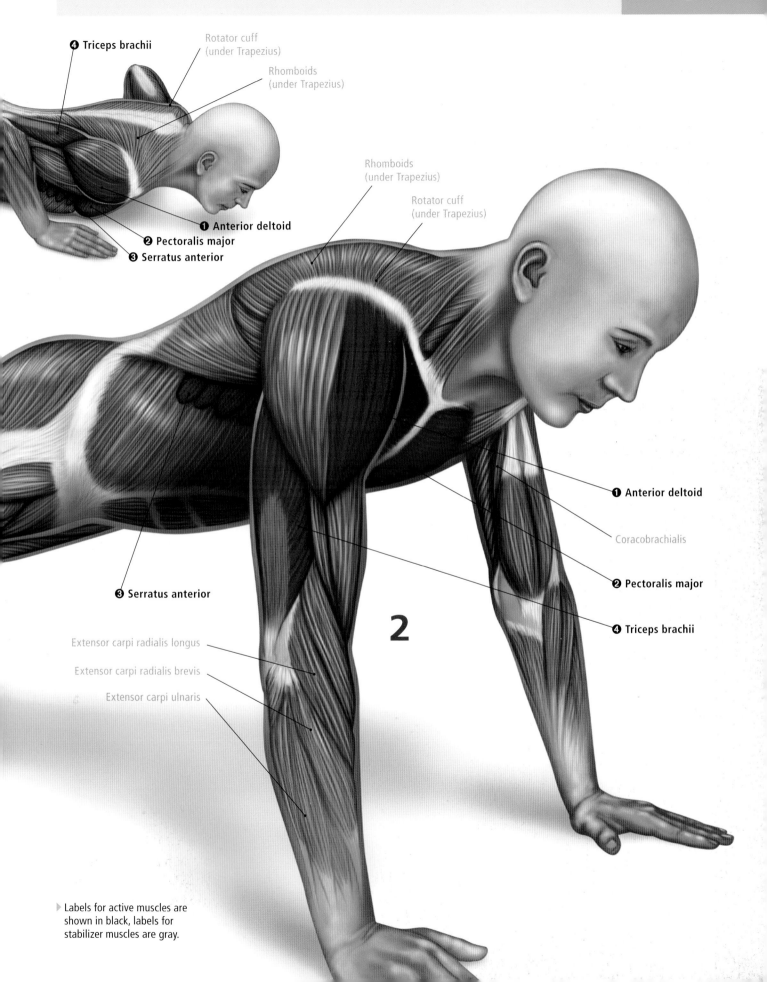

④ Triceps brachii

Rotator cuff
(under Trapezius)

Rhomboids
(under Trapezius)

❶ Anterior deltoid
❷ Pectoralis major
❸ Serratus anterior

Rhomboids
(under Trapezius)

Rotator cuff
(under Trapezius)

❶ Anterior deltoid

Coracobrachialis

❷ Pectoralis major

❹ Triceps brachii

❸ Serratus anterior

2

Extensor carpi radialis longus

Extensor carpi radialis brevis

Extensor carpi ulnaris

▶ Labels for active muscles are
shown in black, labels for
stabilizer muscles are gray.

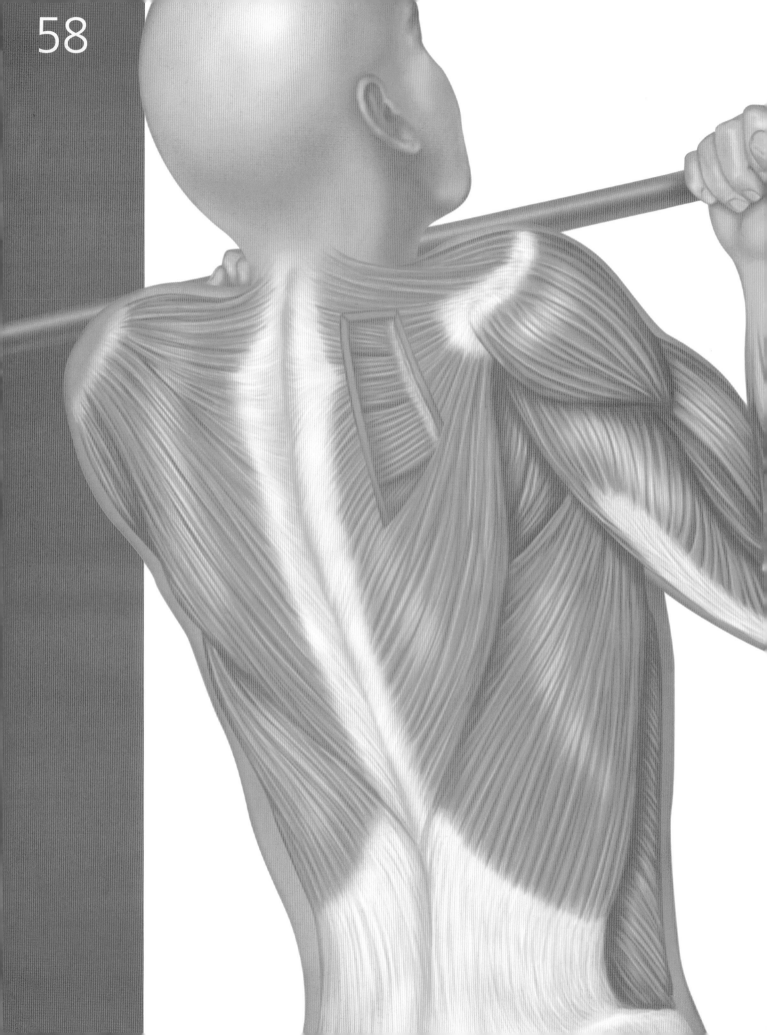

Back Exercises

As the back muscles are not seen in the mirror, they can be neglected during workouts in favor of chest, shoulder, and arm exercises. Do not make this mistake. Back muscles generate the force for pulling and lifting movements, support and protect the spine, and help control the scapula. Training these muscles assists in the performance of other exercises and a number of sporting activities, improves posture, and helps prevent injuries caused by unbalanced training programs.

Choose an exercise combination that targets both the upper and lower back. For beginners, the goal is to learn the correct technique using basic lifts. For those with a developed technique and strength base, increase the variety, volume, and intensity of the weight-training regime.

Lat Pulldown

This well-known exercise is ideal for improving strength in both beginners and advanced exercisers. It targets the muscles of the back, shoulders, and arms, particularly the latissimus dorsi and the biceps, while the hip flexors and abdominal muscles work to stabilize the exerciser on the bench. Other back muscles are also called on to control and retract the scapula. The lat pulldown may improve posture and is recommended for sports that involve gripping and pulling movements. This is a gym-based exercise and should only be performed on the appropriate machine.

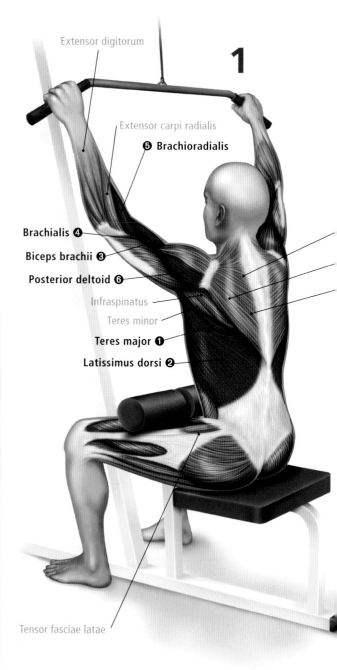

Extensor digitorum

Extensor carpi radialis

❺ Brachioradialis

1

Brachialis ❹

Biceps brachii ❸

Posterior deltoid ❻

Infraspinatus

Teres minor

Teres major ❶

Latissimus dorsi ❷

Tensor fasciae latae

how to

Place hands on the bar just wider than shoulder-width apart, palms facing away from the body. Sit upright, with thighs firmly secured under the padding. Pull the bar down until it is under the chin, squeezing shoulder blades together. Lower the weight in one smooth, controlled motion until arms are straight.

variations

EASY

Try a "close grip" lat pulldown if having trouble keeping a grip on the bar. Grasp the bar with hands just narrower than shoulder-width apart and palms facing toward the body. Pull the bar down until it is under the chin, squeezing shoulder blades together. Lower the weight in one smooth, controlled motion until arms are straight.

HARD

Replace the bar attachment with a rope attachment. This gives the arms and the back a great workout, while really challenging grip.

active muscles

❶ Teres major
❷ Latissimus dorsi
❸ Biceps brachii
❹ Brachialis
❺ Brachioradialis
❻ Posterior deltoid

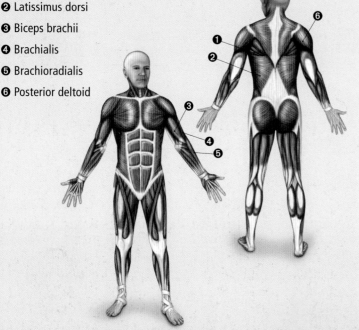

do it right

Maintain an upright posture during the whole exercise and avoid leaning back.

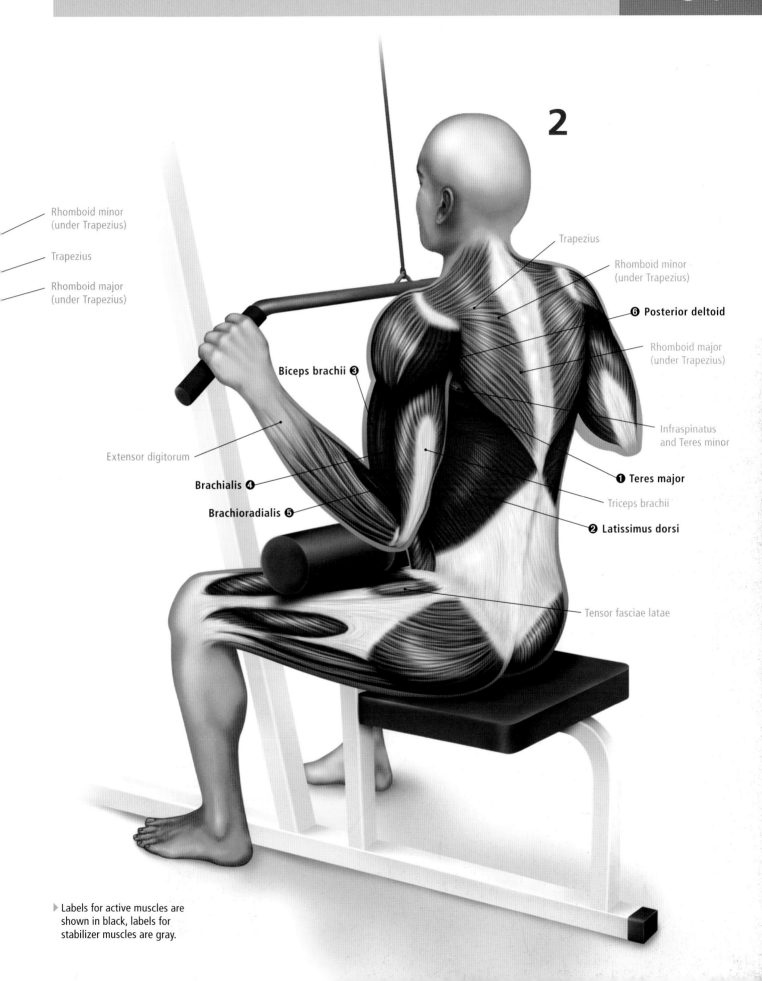

2

Rhomboid minor
(under Trapezius)

Trapezius

Rhomboid major
(under Trapezius)

Trapezius

Rhomboid minor
(under Trapezius)

❻ **Posterior deltoid**

Rhomboid major
(under Trapezius)

Infraspinatus
and Teres minor

Biceps brachii ❸

Extensor digitorum

Brachialis ❹

Brachioradialis ❺

❶ **Teres major**

Triceps brachii

❷ **Latissimus dorsi**

Tensor fasciae latae

▶ Labels for active muscles are
shown in black, labels for
stabilizer muscles are gray.

Chin-up

This strength-building exercise is one that many people find difficult, as the exerciser must lift his or her own bodyweight. Chin-ups develop the back, shoulder, and arm muscles, particularly the latissimus dorsi and the biceps, and build grip strength in the fingers, hands, and forearms. Abdominal muscles are also given a good workout because of the stabilization that is needed through the entire core. This exercise is recommended for any sport that involves gripping, grappling, and pulling, such as martial arts or rock climbing. It requires a stable bar in a gym or on an outdoor climbing frame, or can be performed on doorway bars installed in the home.

warning

Dropping back to the starting position suddenly can lead to hyperextension of the elbows and dislocation of the shoulder joints.

how to

Start with hands on the bar, a shoulder-width apart, with palms facing the body. Hang with knees slightly bent and head upright. In one smooth movement, pull body up until the chin is above the height of the bar, then gently lower body to the starting position. Ensure arms are fully extended when the chin-up is completed.

variations

EASY

Use a spotter to build strength if new to the exercise and finding it difficult to maintain correct technique. While in the starting position, bend at the knees so that the spotter can support the ankles. If necessary, push against this support base while raising body up toward the bar.

HARD

Try a pull-up to target the middle back muscles. The pull-up is performed in exactly the same way as the chin-up, except that hands are placed on the bar with palms facing away from the body. This technique is a favorite of the military and emergency services.

active muscles

❶ Trapezius
❷ Posterior deltoid
❸ Teres minor
❹ Teres major
❺ Biceps brachii
❻ Brachialis
❼ Brachioradialis
❽ Latissimus dorsi
❾ Rhomboid major
❿ Rhomboid minor

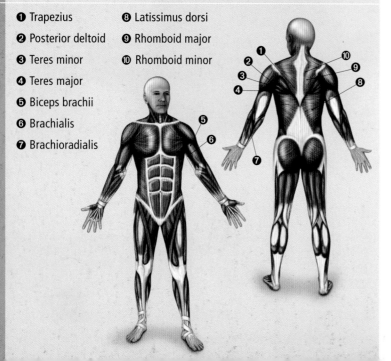

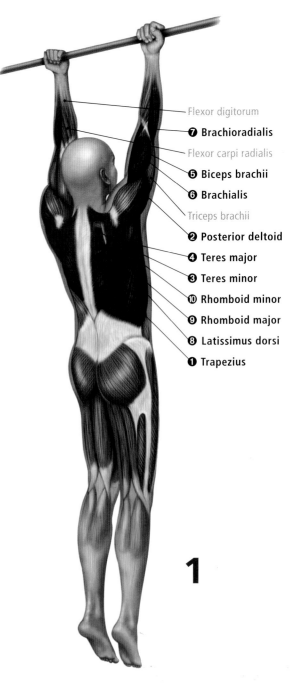

Flexor digitorum
❼ Brachioradialis
Flexor carpi radialis
❺ Biceps brachii
❻ Brachialis
Triceps brachii
❷ Posterior deltoid
❹ Teres major
❸ Teres minor
❿ Rhomboid minor
❾ Rhomboid major
❽ Latissimus dorsi
❶ Trapezius

1

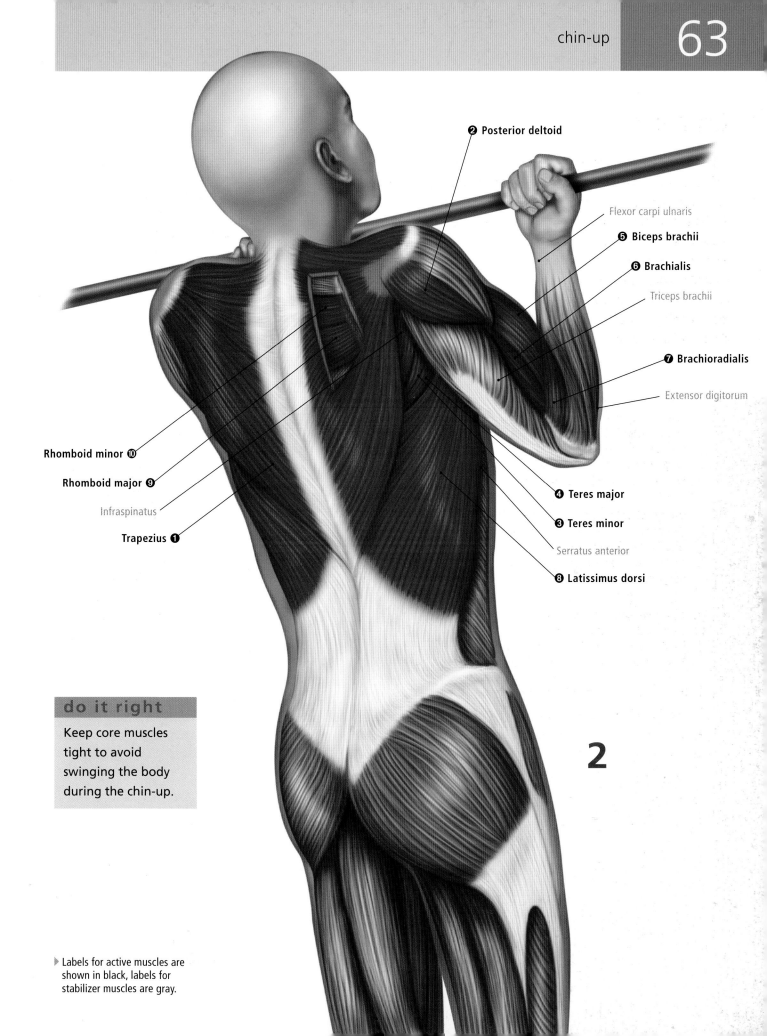

❷ Posterior deltoid

Flexor carpi ulnaris

❺ Biceps brachii

❻ Brachialis

Triceps brachii

❼ Brachioradialis

Extensor digitorum

Rhomboid minor ❿

Rhomboid major ❾

Infraspinatus

Trapezius ❶

❹ Teres major

❸ Teres minor

Serratus anterior

❽ Latissimus dorsi

do it right
Keep core muscles tight to avoid swinging the body during the chin-up.

2

▶ Labels for active muscles are shown in black, labels for stabilizer muscles are gray.

Bent-over Row

1

This free-weight exercise targets the muscles of the upper back, while calling on the lower back and leg muscles to provide support. In particular, the latissimus dorsi, rear deltoid, infraspinatus, and biceps are targeted, while the erector spinae and hamstring muscles must contract strongly to support the upper body. The bent-over row is suitable for intermediate or advanced exercisers and is ideal for sports and occupations that require bending, lifting, or pulling. It can be performed in either a gym or home setting, as it requires only some free floor space and a bar or dumbbells.

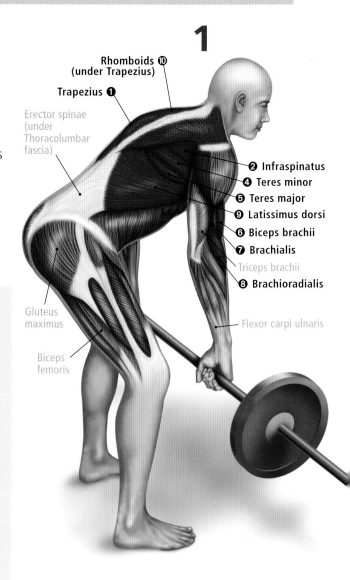

Rhomboids ❿
(under Trapezius)

Trapezius ❶

Erector spinae
(under
Thoracolumbar
fascia)

❷ Infraspinatus
❹ Teres minor
❺ Teres major
❾ Latissimus dorsi
❻ Biceps brachii
❼ Brachialis
Triceps brachii
❽ Brachioradialis

Gluteus
maximus

Flexor carpi ulnaris

Biceps
femoris

how to

Grip the bar with hands just wider than shoulder-width apart, palms facing toward the body. Feet should be shoulder-width apart and knees slightly bent. In the starting position, bend forward at the hips so the bar is just below the knees, with back straight. Pull bar up to the ribs, while keeping torso still and elbows close to sides. Lower again in a smooth, controlled movement.

variations

EASY

Use two dumbbells and grasp weights with palms facing each other. This makes the exercise easier on the grip and allows the weight to move, without the knees getting in the way. Pull the dumbbells up to the ribs, keeping elbows close to sides. Lower again in a smooth, controlled movement.

HARD

Perform the row with hands shoulder-width apart and palms facing away from the body to target the latissimus dorsi and provide an extra challenge for the biceps. Pull the bar up to the ribs, keeping elbows close to sides. Lower again slowly until arms are fully extended.

active muscles

❶ Trapezius
❷ Infraspinatus
❸ Posterior deltoid
❹ Teres minor
❺ Teres major
❻ Biceps brachii
❼ Brachialis
❽ Brachioradialis
❾ Latissimus dorsi
❿ Rhomboids
(under Trapezius)

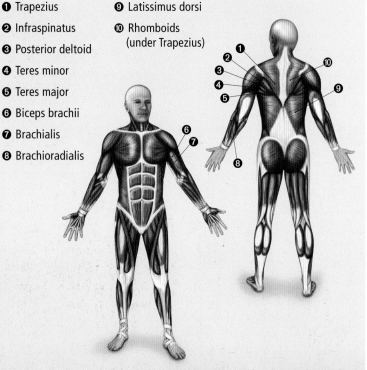

warning

Only bend forward at the waist as far as flexibility allows. Do not round the back during this exercise.

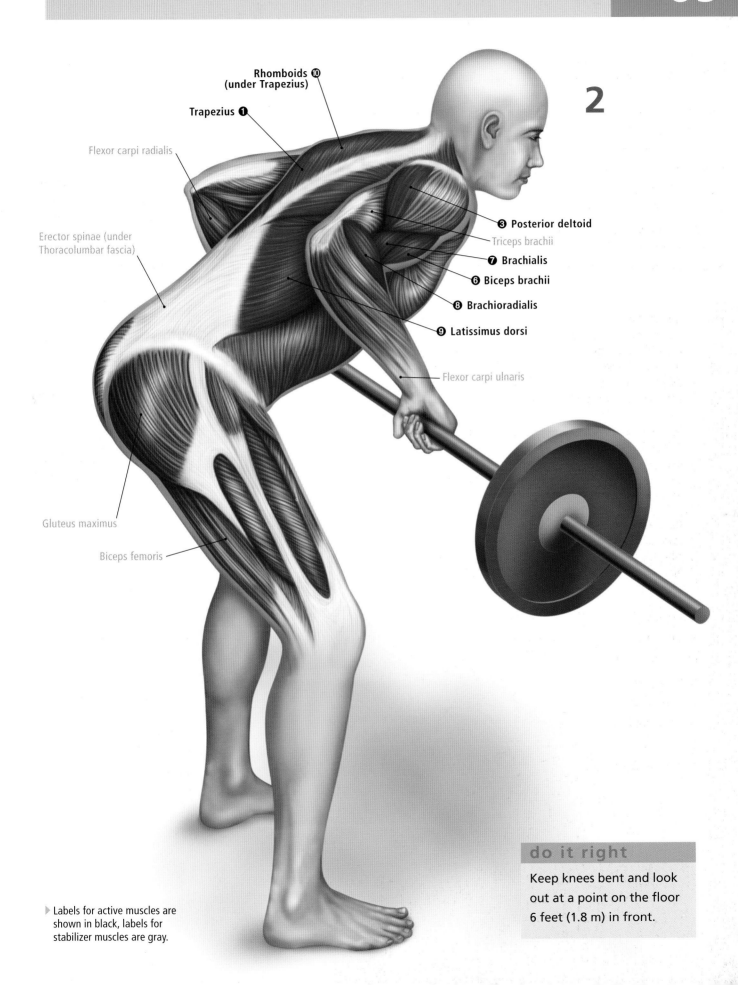

2

Rhomboids ⑩
(under Trapezius)

Trapezius ❶

Flexor carpi radialis

❸ **Posterior deltoid**

Triceps brachii

❼ **Brachialis**

❻ **Biceps brachii**

❽ **Brachioradialis**

❾ **Latissimus dorsi**

Erector spinae (under
Thoracolumbar fascia)

Flexor carpi ulnaris

Gluteus maximus

Biceps femoris

▷ Labels for active muscles are
shown in black, labels for
stabilizer muscles are gray.

Seated Row

This strengthening exercise is performed in a gym setting. It is highly effective in providing a workout for the whole back. The muscles that control the scapula are given extra attention, with the latissimus dorsi, rear deltoids, and biceps bearing the brunt of the load. The leg, gluteal, and lower back muscles are also required to provide a stable support base, bracing in the sitting position. Use the seated row to add some postural balance back into a workout routine because the muscles that retract and depress the scapula are often weak and underutilized. This exercise requires an appropriate seated-row machine or a low cable pulley. It can be performed by both beginners and advanced trainers.

1

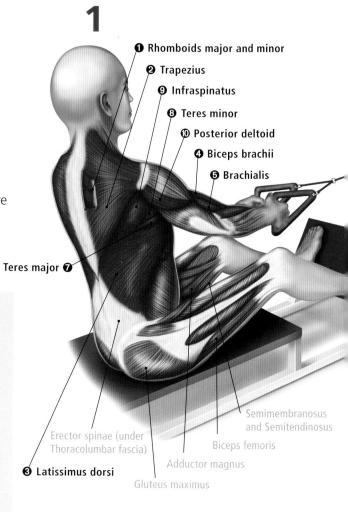

❶ Rhomboids major and minor
❷ Trapezius
❾ Infraspinatus
❽ Teres minor
❿ Posterior deltoid
❹ Biceps brachii
❺ Brachialis
Teres major ❼
❸ Latissimus dorsi
Erector spinae (under Thoracolumbar fascia)
Gluteus maximus
Adductor magnus
Biceps femoris
Semimembranosus and Semitendinosus

how to

Most machines have a handle where the hands are close together. Grasp this with palms facing toward each other. Place feet on the footholds and bend knees at about 30 degrees. Start with arms fully extended, then squeeze shoulder blades together and pull the bar toward the ribs, so elbows brush by sides. Allow the weight to extend slowly back to the starting position.

variations

EASY Use a machine that has a chest support, which allows the exercise to be performed without placing any demand on the lower back or legs. However, progress to the standard version of the seated row once the easy variation has been mastered.

HARD Use a long bar with hands just wider than shoulder-width apart, palms facing down. This reduces the load on the latissimus dorsi and targets the rear deltoids and rhomboid muscles. Squeeze shoulder blades together and pull the bar toward the chest, keeping elbows level with the bar.

active muscles

❶ Rhomboids major and minor
❷ Trapezius
❸ Latissimus dorsi
❹ Biceps brachii
❺ Brachialis
❻ Brachioradialis
❼ Teres major
❽ Teres minor
❾ Infraspinatus
❿ Posterior deltoid

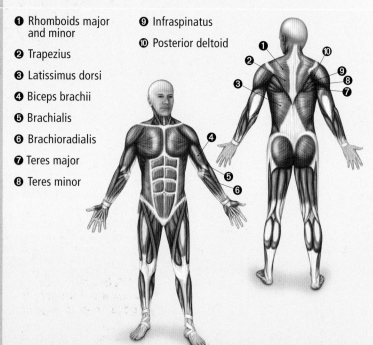

warning

Leaning forward or backward during the movement places unnecessary stress on the lower back.

2

do it right

Maintain a tall sitting position throughout the whole movement.

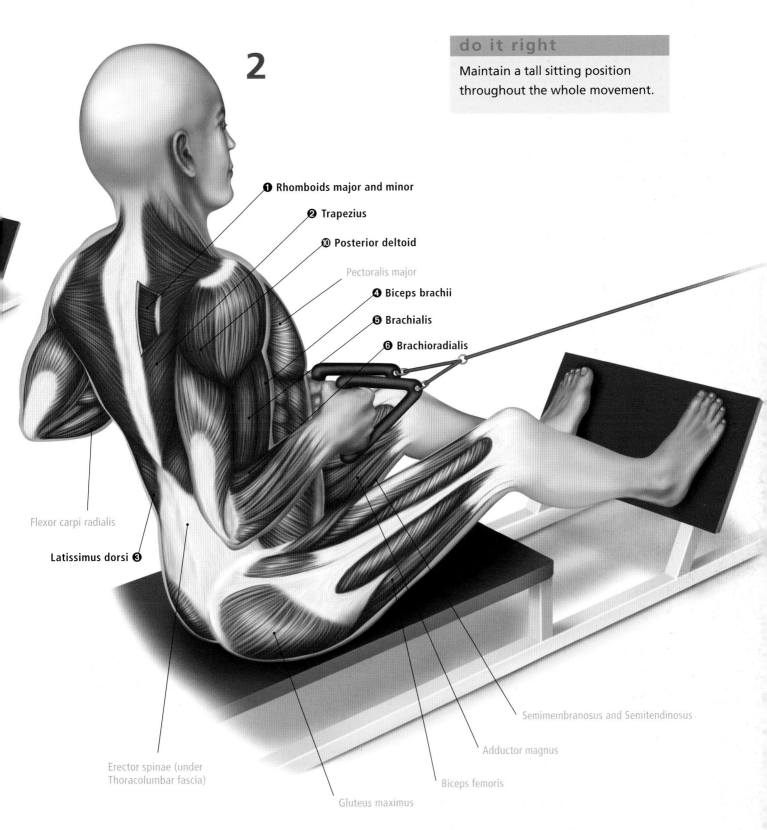

❶ Rhomboids major and minor

❷ Trapezius

❿ Posterior deltoid

Pectoralis major

❹ Biceps brachii

❺ Brachialis

❻ Brachioradialis

Flexor carpi radialis

Latissimus dorsi ❸

Semimembranosus and Semitendinosus

Adductor magnus

Biceps femoris

Erector spinae (under Thoracolumbar fascia)

Gluteus maximus

▶ Labels for active muscles are shown in black, labels for stabilizer muscles are gray.

Reverse Fly

This simple-but-effective exercise targets the muscles of the upper back and shoulders. The rear deltoid does most of the heavy lifting, assisted by the infraspinatus, teres minor, trapezius, and rhomboids. Recommended for both beginners and advanced exercisers, the reverse fly should be scheduled at the end of a workout, after the pulling exercises, to make sure these muscles have been worked hard enough during the session. This exercise can help prevent common shoulder injuries experienced in racket sports because it strengthens the muscles around the shoulder. Two dumbbells, a bit of space, and a sturdy bench or chair to sit on are all that is needed to perform the reverse fly.

warning

At the top of the movement, elbows should be at right angles to the torso to avoid placing stress on the rotator cuff muscles.

how to

Sit at the end of a chair or bench and lean forward so the torso rests on the thighs. Grasp the dumbbells either underneath the legs or just next to the feet, with palms facing toward each other. Raise arms to the side, squeezing shoulder blades together, until elbows are shoulder height. Slowly lower back into the starting position, being careful that dumbbells do not collide with the ankles.

variations

EASY

Lie facedown on a bench to perform the reverse fly if unable to get into a comfortable sitting position. Ensure that the back does not arch during the lift.

HARD

Perform the exercise in the standing position, with feet hip-width apart and knees slightly bent. Lean forward at the hips, keeping the back straight. Replace dumbbells with a low pulley cable, or for an even greater challenge, exercise one arm at a time.

active muscles

❶ Posterior deltoid

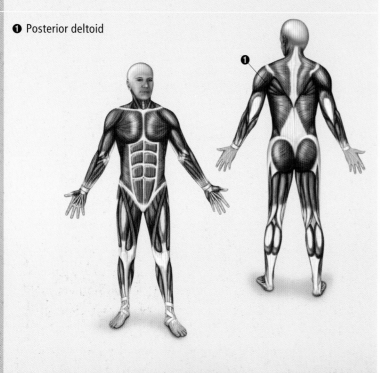

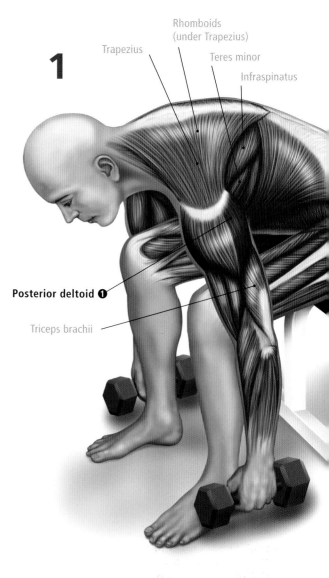

1

Rhomboids (under Trapezius)

Trapezius

Teres minor

Infraspinatus

Posterior deltoid ❶

Triceps brachii

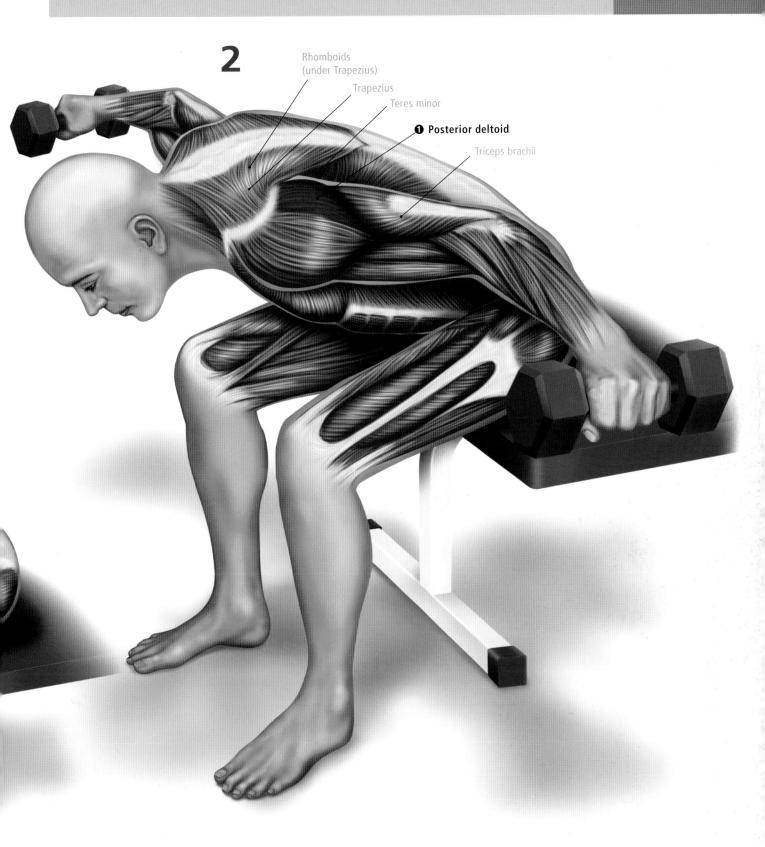

2

Rhomboids
(under Trapezius)

Trapezius

Teres minor

❶ **Posterior deltoid**

Triceps brachii

▶ Labels for active muscles are
shown in black, labels for
stabilizer muscles are gray.

do it right

Keep elbows slightly bent
throughout this exercise.

Single-arm Row

This exercise provides the back muscles with a great workout in a supported, stable position. The latissimus dorsi, rear deltoids, and biceps muscles are targeted, while the muscles around the scapula, forearm, and grip strength are also developed. The single-arm row is performed in a position that supports the lower back, so it suits both beginners and advanced exercisers. It is ideal for any sports that involve gripping and pulling, such as contact sports, rowing, and kayaking, and is a powerful exercise for increasing strength. All that is required is a sturdy bench at approximately knee height and one dumbbell.

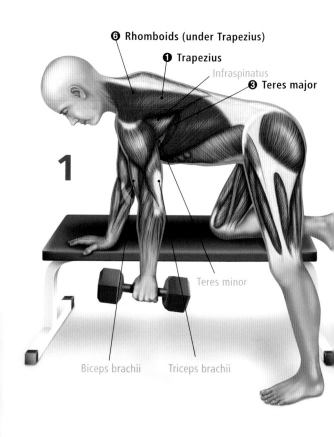

❻ Rhomboids (under Trapezius)
❶ Trapezius
Infraspinatus
❸ Teres major
1
Teres minor
Biceps brachii   Triceps brachii

how to

Place knee and hand of the supporting arm on a bench, so torso is horizontal. Position the other foot on the ground, slightly back and to the side for stability. Grasp the dumbbell with palm facing toward the bench. Pull the dumbbell up toward the body until it is at torso height, with elbow brushing by the side of the body. Slowly lower the weight until the arm is fully extended.

variations

EASY Lie facedown on the bench if uncomfortable or feeling unstable in the kneeling position. Ensure the back does not arch during the lift.

HARD For an extra challenge, perform the exercise in a standing position, with feet shoulder-width apart and knees bent. Try using a low pulley cable without any support for the opposite hand.

active muscles

❶ Trapezius
❷ Posterior deltoid
❸ Teres major
❹ Brachialis
❺ Latissimus dorsi
❻ Rhomboids (under Trapezius)

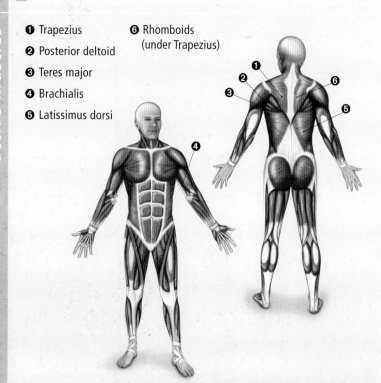

do it right

Keep back straight during the lift and look at a spot on the ground 4 feet (1.2 m) in front.

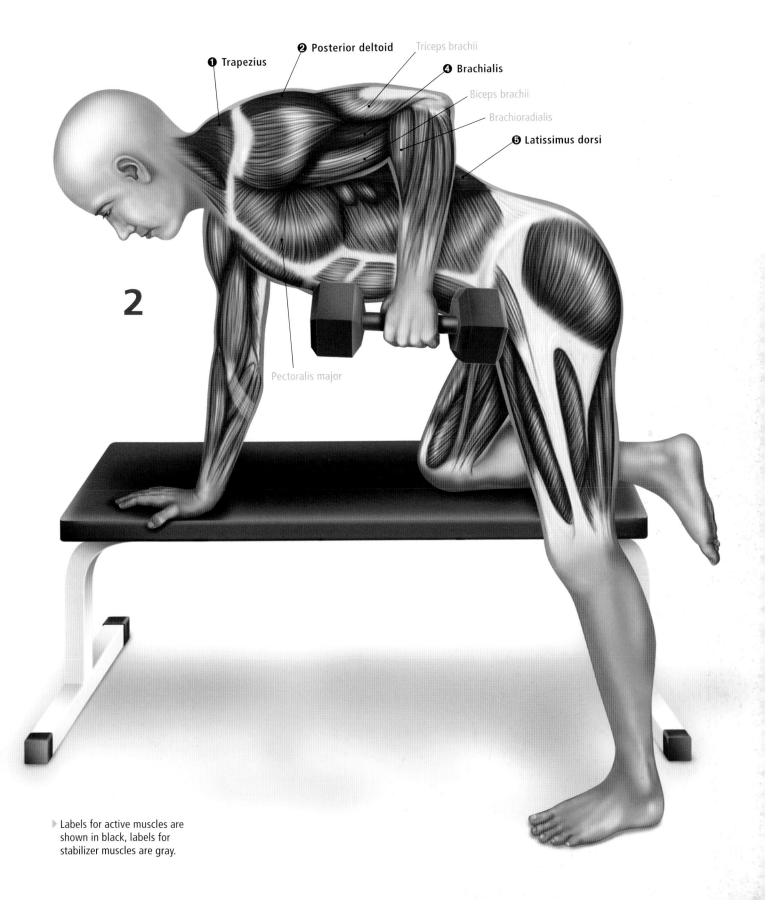

❶ Trapezius
❷ Posterior deltoid
Triceps brachii
❹ Brachialis
Biceps brachii
Brachioradialis
❺ Latissimus dorsi

2

Pectoralis major

▷ Labels for active muscles are shown in black, labels for stabilizer muscles are gray.

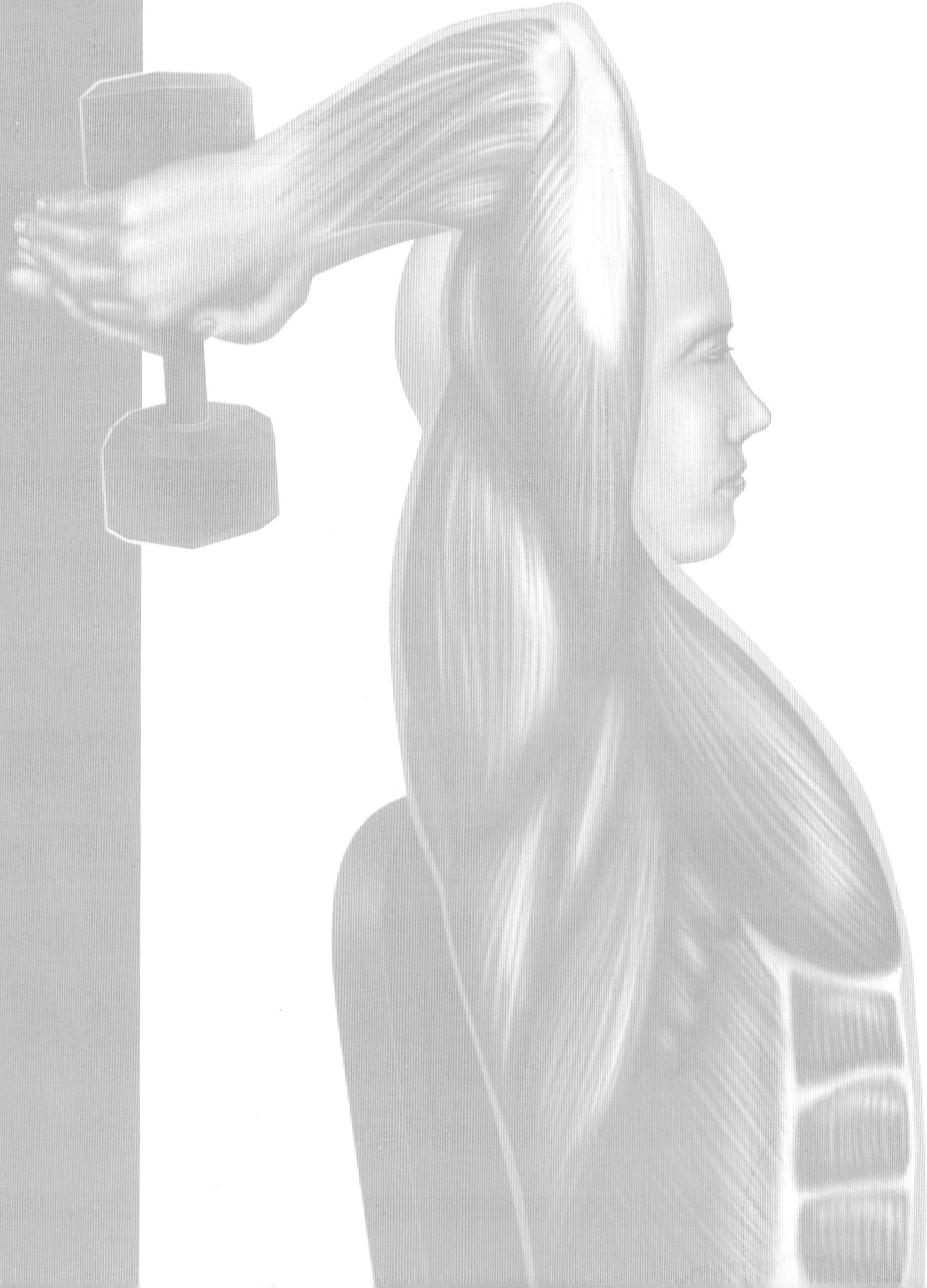

Arm and Shoulder Exercises

The deltoid is the target of shoulder exercises in this section. Its three sets of fibers generate force, while the rotator cuff muscles provide dynamic stability and allow for the range of motion around the shoulder (glenohumeral) joint. Strong, stable arms and shoulders enhance performance in many sports. Arm exercises address the muscles that flex the elbow—biceps brachii, brachialis, and brachioradialis—as well as triceps brachii, which extends the elbow.

Both shoulder and arm muscles are engaged during exercises for other muscle groups. To avoid fatigue and overuse injuries, allow adequate rest and recovery between workouts. The anterior deltoid and rotator cuff muscles are particularly susceptible, so it is important to pay attention to the posterior deltoid.

Biceps Curl

This classic exercise has been a staple of strength-training regimens for decades. It is a simple exercise that can be performed by both beginners and advanced trainers, and has many variations. The biceps curl targets the biceps brachii, but also gives other muscles of the upper arm, brachialis, and brachioradialis a good workout. The muscles of the shoulder and forearm must also contract to support the movement this exercise demands. Requiring only some floor space and a barbell or pair of dumbbells, this exercise can be performed almost anywhere.

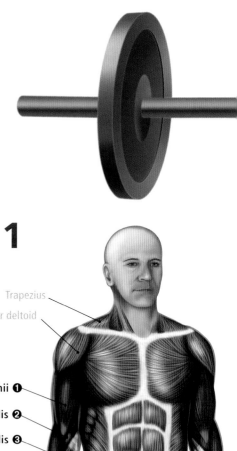

1

how to

Stand with feet shoulder-width apart and knees slightly bent. Grasp the bar with hands shoulder-width apart, palms facing away from the body. Move hands closer or further apart if these positions are more comfortable. Raise the bar until forearms are vertical, keeping elbows fixed by sides. Lower the bar until arms are fully extended.

variations

EASY Stand with one foot forward and one foot back for extra stability if feeling unsteady or lifting heavy loads. This position also helps keep the torso still and provides extra support for the lower back.

HARD Sit or stand with one dumbbell in each hand. Start with palms facing toward each other, then with each dumbbell curl, rotate wrists so that palms face the shoulder at the top of the movement.

active muscles

❶ Biceps brachii
❷ Brachialis
❸ Brachioradialis

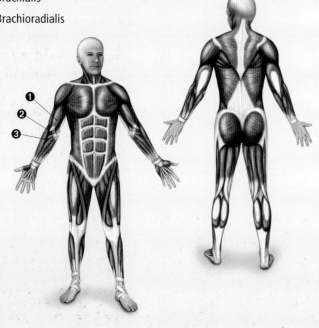

Trapezius
Anterior deltoid
Biceps brachii ❶
Brachialis ❷
Brachioradialis ❸
Flexor carpi radialis
Flexor carpi ulnaris

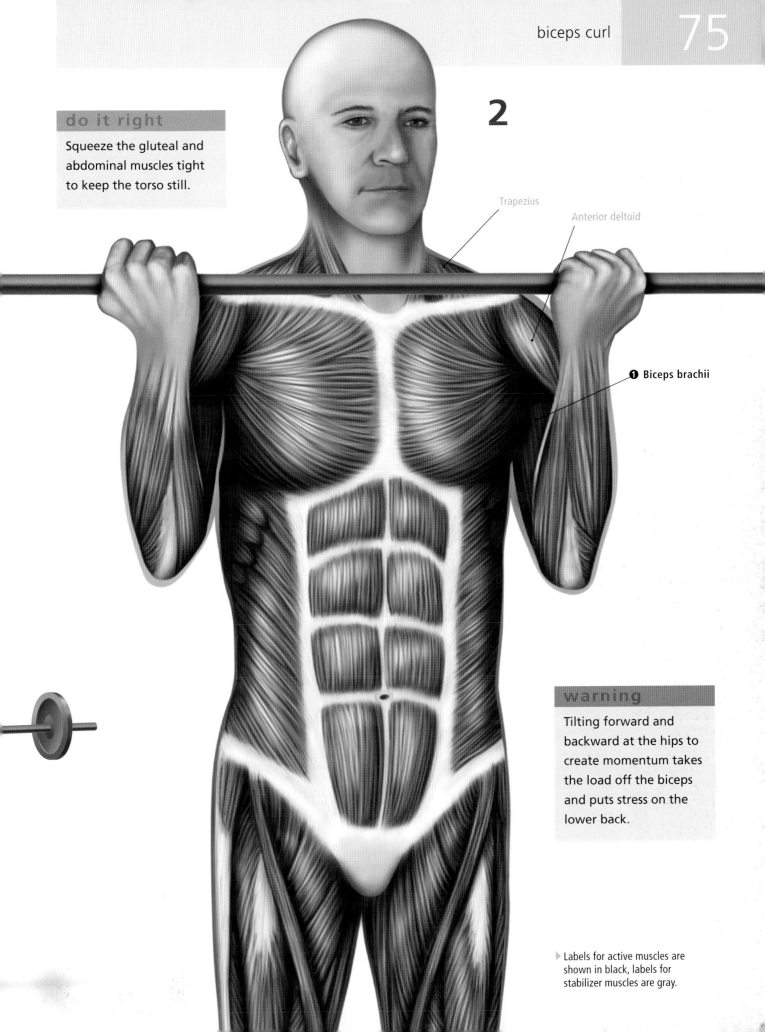

2

do it right

Squeeze the gluteal and abdominal muscles tight to keep the torso still.

Trapezius

Anterior deltoid

❶ Biceps brachii

warning

Tilting forward and backward at the hips to create momentum takes the load off the biceps and puts stress on the lower back.

▶ Labels for active muscles are shown in black, labels for stabilizer muscles are gray.

Concentration Curl

This simple exercise targets the muscles of the upper arm. Brachialis is the most active of these muscles, as shoulder flexion disadvantages the short head of biceps brachii. This focus makes the concentration curl a good addition, or at least a good occasional alternative, to more traditional biceps exercises. Include it in an exercise program to ensure maximum upper arm workout. The stable seated position reduces stress on the lower back, limits the use of momentum, and allows the muscles of the upper arm to be isolated.

warning

Avoid twisting, rotating, or lifting the torso in order to lift the weight. If any of these actions are required, the weight is too heavy.

do it right

Place other hand on the other thigh to support the upper body.

how to

Sit on a bench with the dumbbell between the feet. Lean forward slightly so that the elbow rests against the inner thigh. Grasp the dumbbell with palm facing away from the body, then curl the arm up until the palm faces the shoulder. Stabilize the arm with the thigh, so the elbow remains still throughout the exercise. Lower the weight until the arm is fully extended.

variations

EASY Place the elbow on top of the thigh rather than against the inner thigh, to allow for a little extra leverage and to make the exercise easier at the bottom of the lift.

HARD Perform a "hammer curl" by rotating the wrist during the curl, so the thumb faces the ceiling at the top of the movement, with palm facing inward. This variation targets the other elbow flexor muscle, brachioradialis.

active muscles

❶ Brachialis
❷ Biceps brachii
❸ Brachioradialis

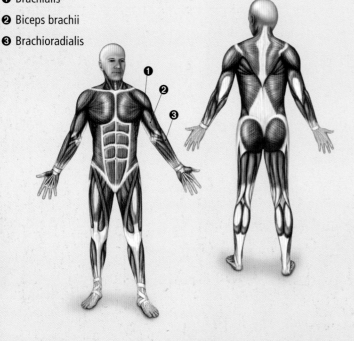

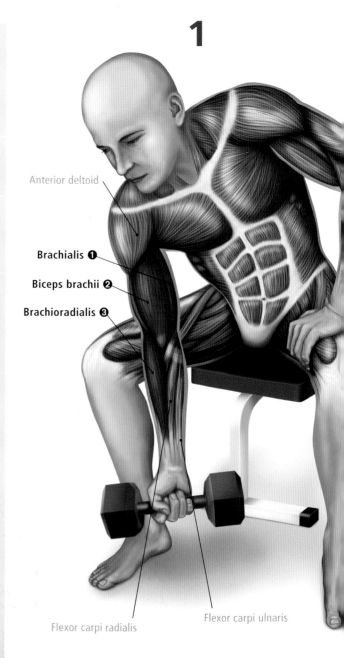

1

Anterior deltoid

Brachialis ❶
Biceps brachii ❷
Brachioradialis ❸

Flexor carpi radialis

Flexor carpi ulnaris

2

Anterior deltoid

Biceps brachii ❷

Brachioradialis ❸

▶ Labels for active muscles are shown in black, labels for stabilizer muscles are gray.

Cable Curl

This biceps exercise is commonly performed at the end of a workout using lighter weights. Brachialis and the long head of biceps brachii are targeted, although as with any exercise in the standing position, the abdominal and lower back muscles are engaged to stabilize the torso, while the rotator cuff muscles must contract to support and stabilize the shoulder. The cable curl is a good exercise for throwing sports, such as baseball, cricket, and water polo, because the biceps muscles are strengthened in a position where they contribute to shoulder stability in the overhead throwing position. This exercise requires two high pulley cables, which makes it best suited to a gym setting.

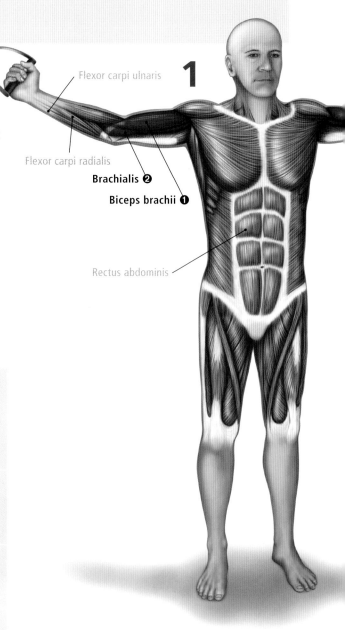

Flexor carpi ulnaris

Flexor carpi radialis

Brachialis ❷

Biceps brachii ❶

Rectus abdominis

1

how to

Using two high pulley cables, grasp each where the cables are far enough apart to stand in the center with arms outstretched. Palms face toward the ceiling. Curl handles toward the shoulders, keeping upper arms and body still. Slowly return to the starting position until arms are fully extended.

variations

EASY

Grasp the handles of the cables, then take a step backward so that rather than being straight out to the side, arms are angled forward at 45 degrees. This takes the stress off the shoulders. Curl handles toward the shoulders, keeping upper arms and body still.

HARD

Use a rope attachment on each pulley for an extra grip challenge. Grasp the rope so that palms face away from the body, then rotate wrists during the lift so that palms face toward the shoulders.

active muscles

❶ Biceps brachii
❷ Brachialis

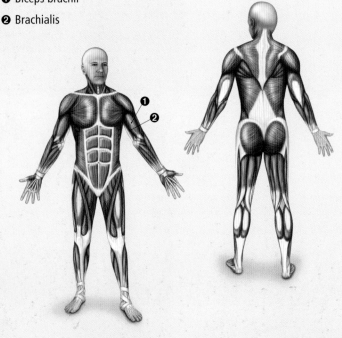

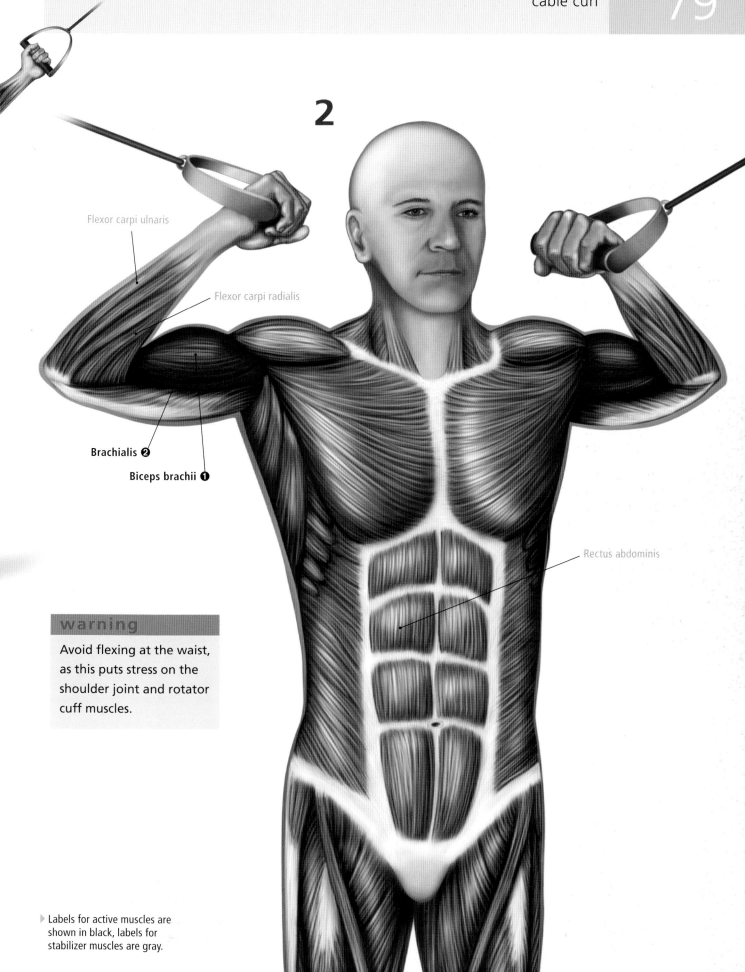

2

Flexor carpi ulnaris

Flexor carpi radialis

Brachialis ❷

Biceps brachii ❶

Rectus abdominis

warning

Avoid flexing at the waist, as this puts stress on the shoulder joint and rotator cuff muscles.

▶ Labels for active muscles are shown in black, labels for stabilizer muscles are gray.

Triceps Pushdown

This strength-building exercise is a common sight in gyms and fitness centers, performed by people looking to improve the size and strength of their triceps muscles. The versatile pushdown targets the triceps brachii and can be performed with many variations, using different grips and cable attachments. Abdominal muscles get a great secondary workout from this exercise, especially as strength increases and trainers start to lift heavier weights. Suitable for both beginners and the more advanced, the triceps pushdown is best performed in a gym.

1

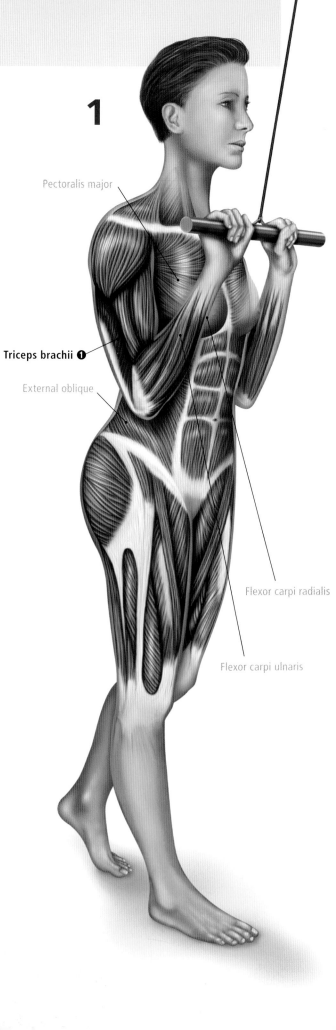

Pectoralis major

Triceps brachii ❶

External oblique

Flexor carpi radialis

Flexor carpi ulnaris

how to

Facing a high pulley cable, grasp cable attachment with hands spaced just narrower than shoulder-width apart and palms facing the floor. Extend arms down, pushing the bar toward the floor while keeping elbows by sides and the abdominal muscles tight. Bend at the elbows and return the bar until the forearms are close to the upper arms.

variations

EASY

Stand with one foot forward and one foot back for extra stability if feeling unsteady or lifting heavy loads. This also helps keep the torso still. It may be more comfortable to use a bar that slopes downward at 45 degrees on each side, as this takes pressure off the wrists.

HARD

Use a rope attachment on the pulley and grip the rope with thumbs facing the ceiling. Keep hands close together at the top of the movement, rotating the wrists so that palms face the floor at the bottom of the movement. This targets more of the triceps muscle and provides a real grip challenge.

active muscles

❶ Triceps brachii

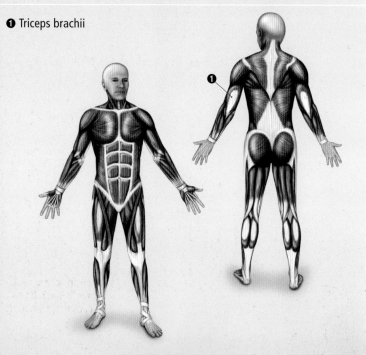

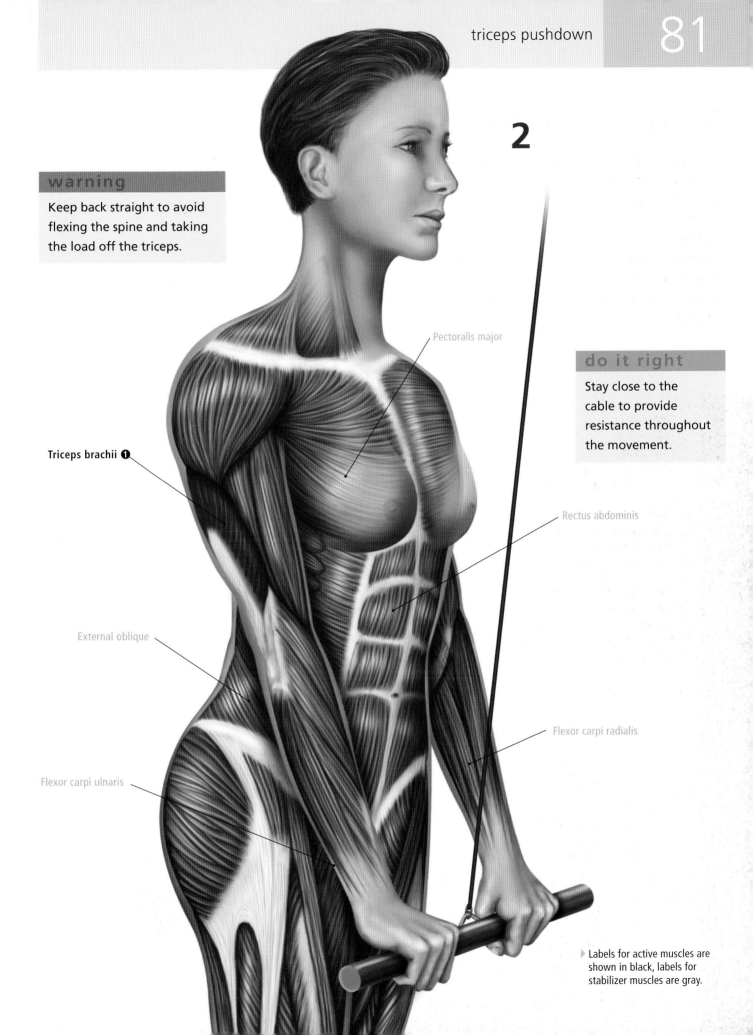

2

Pectoralis major

Triceps brachii ❶

Rectus abdominis

External oblique

Flexor carpi radialis

Flexor carpi ulnaris

▶ Labels for active muscles are shown in black, labels for stabilizer muscles are gray.

Triceps Extension

This strengthening exercise uses a single dumbbell to effectively target the triceps brachii muscle. The wrist flexor and deltoid help to stabilize the arm, while the triceps works from its most lengthened position, stretching at the bottom of the movement. Anyone training for sports that require strength in overhead positions, such as serving or smashing movements in tennis or volleyball, or in throwing sports, may prefer triceps extensions over triceps pushdowns. This exercise can be performed at home or in a gym, although it requires an attentive spotter to help control the weight, especially when lifting over the head.

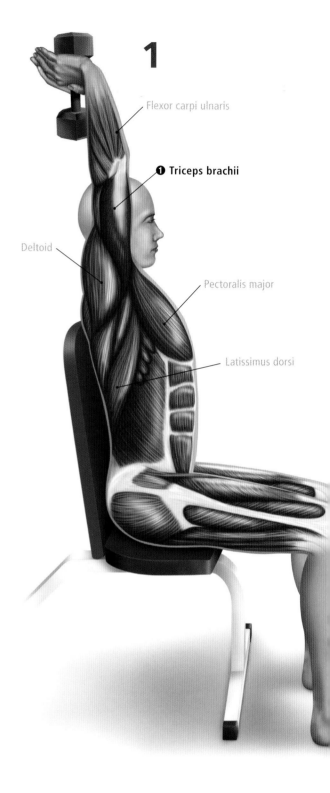

1

Flexor carpi ulnaris

❶ **Triceps brachii**

Deltoid

Pectoralis major

Latissimus dorsi

how to

Hold a dumbbell with both palms against the plate on one side and thumbs wrapped around the handle. Sitting in a chair or on a bench with back support, start with the dumbbell overhead, arms fully extended. Keep elbows still and lower the weight behind the head by flexing the elbows. Raise dumbbell overhead by extending the elbows.

variations

EASY

Lie back on a bench and grasp one dumbbell in each hand, palms facing each other. Start with arms fully extended toward the ceiling and lower the dumbbells down beside the head, keeping elbows fixed in place. Raise the dumbbells toward the ceiling until arms are again fully extended.

HARD

Perform a one-arm extension by holding the dumbbell with one hand, palm facing forward. In a sitting position, start with the dumbbell overhead, with arm fully extended. Keep elbow still and lower the weight behind the head by flexing the elbow. Raise the dumbbell overhead by fully extending the elbow.

active muscles

❶ Triceps brachii

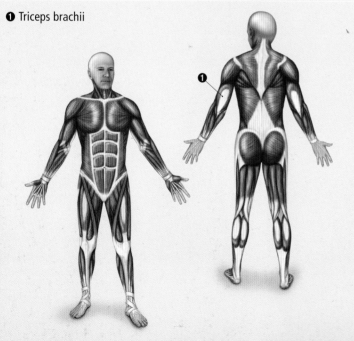

warning

Lower the weight slowly to avoid hitting the back of the neck.

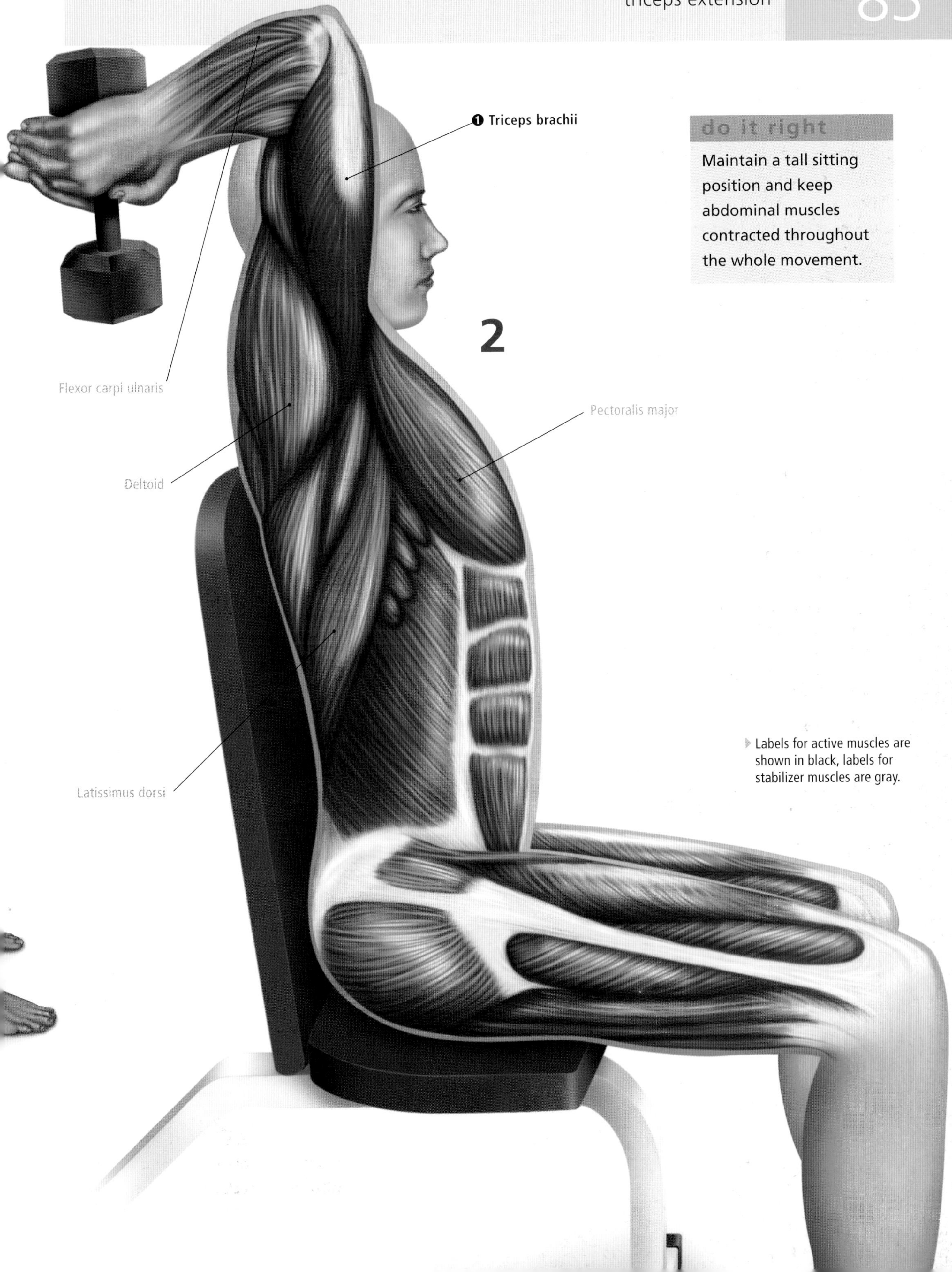

❶ **Triceps brachii**

Flexor carpi ulnaris

Deltoid

Latissimus dorsi

do it right

Maintain a tall sitting position and keep abdominal muscles contracted throughout the whole movement.

2

Pectoralis major

▶ Labels for active muscles are shown in black, labels for stabilizer muscles are gray.

Triceps Kickback

This simple exercise is performed using a bench and a dumbbell. It targets the triceps brachii at the back of the upper arm. Unlike other triceps exercises, the triceps kickback limits the effective range of motion, making it a suitable exercise for beginners and a good occasional addition to the traditional triceps exercises that more advanced trainers perform. The bench supports the torso, which means there is little or no strain on the lower back. The triceps brachii works hardest when in the shortened position, so trainers will really feel the muscle squeezing at the top of the movement.

warning
Do not rotate the torso to swing the weight up.

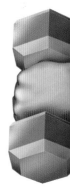

how to

Place knee and hand of the supporting arm on a bench so that the torso is horizontal. Position the other foot on the ground, slightly back and to the side for stability. Grasp the dumbbell with palm facing toward the bench, elbow by the side, and arm bent at 90 degrees. Keeping the elbow fixed at the side, extend arm at the elbow until it points straight back. Lower the dumbbell and return to the starting position.

variations

EASY
If experiencing difficulty, lower the elbow so that it is slightly lower than the body. This reduces the range of motion that the triceps has to work through to complete the movement. However, it also reduces the effectiveness of the exercise.

HARD
Grasp the dumbbell with palm facing forward, elbow by the side and arm bent at 90 degrees. This grip recruits more of the triceps brachii muscle. Keeping the elbow fixed at the side, extend arm at the elbow until it points straight back. Lower the dumbbell and return to the starting position.

active muscles

❶ Triceps brachii

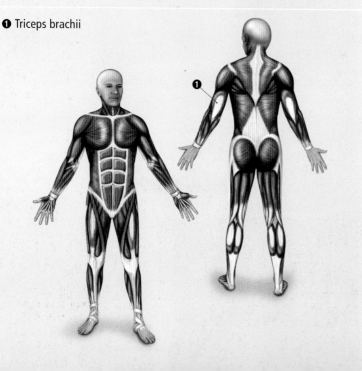

❶ Triceps brachii
Extensor carpi ulnaris
Posterior deltoid

1

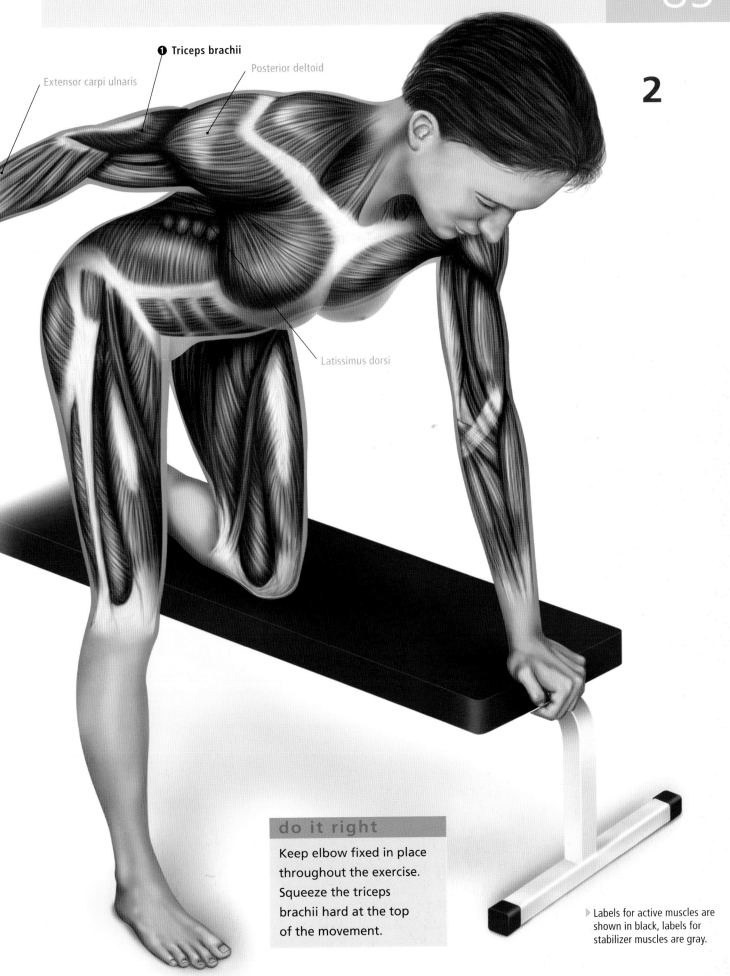

❶ Triceps brachii

Posterior deltoid

Extensor carpi ulnaris

2

Latissimus dorsi

do it right

do it right

Keep elbow fixed in place throughout the exercise. Squeeze the triceps brachii hard at the top of the movement.

▸ Labels for active muscles are shown in black, labels for stabilizer muscles are gray.

Shoulder Press

This exercise, and its many variations, is a staple of any complete program for intermediate and advanced trainers. The shoulder press develops the anterior and lateral deltoids, and provides a good workout for triceps brachii. Lifting dumbbells overhead also requires a great deal of core and shoulder stability, recruiting additional muscles for the task, such as supraspinatus, trapezius, and serratus anterior, the abdominal muscles, and the lower back muscles. This exercise is ideal for sports or occupations that involve pushing or lifting overhead, such as rugby, martial arts, or dancing. It can be performed in a gym or at home but should be done with a training partner to act as a spotter.

warning

Always use a spotter for overhead lifts to prevent injury from dropping weights.

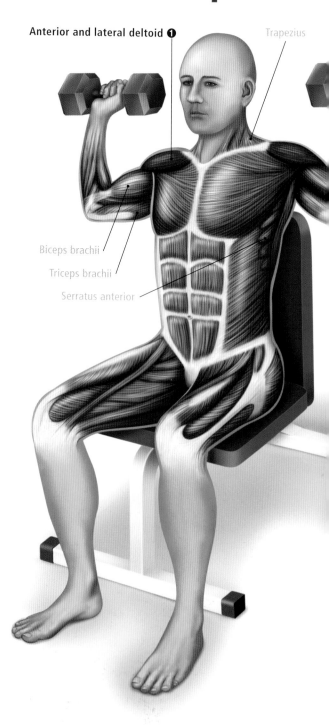

1

Anterior and lateral deltoid ❶

Trapezius

Biceps brachii

Triceps brachii

Serratus anterior

how to

Sit on a bench or chair with back support, feet apart. Grasp each dumbbell, holding them at ear height with palms facing forward and forearms perpendicular to the ground. Press dumbbells upward until arms are fully extended overhead, allowing the dumbbells to come together at the top. Slowly lower the weights until they are in the starting position. When lifting heavy weights, use a spotter or a rack to help get the bar into the correct starting position.

variations

EASY

Perform a seated "military press" on a bench or chair with back support, feet apart. Grasp a barbell with hands slightly wider than shoulder-width apart and palms facing forward. Press the bar upward until arms are fully extended overhead. Slowly lower the bar until it is in the starting position.

HARD

Perform an "Arnold press" by grasping each dumbbell and holding them in front of the shoulders, with palms facing the body and elbows under wrists—like the top position of a chin-up. Press dumbbells upward and bring elbows out to the sides until arms are fully extended overhead, rotating so palms face forward. Slowly lower the weights, rotating again until they are in the starting position.

active muscles

❶ Anterior and lateral deltoid

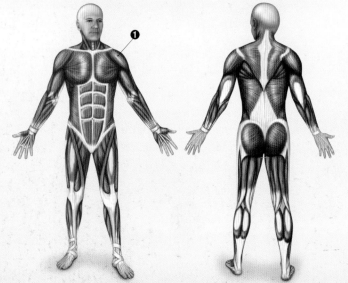

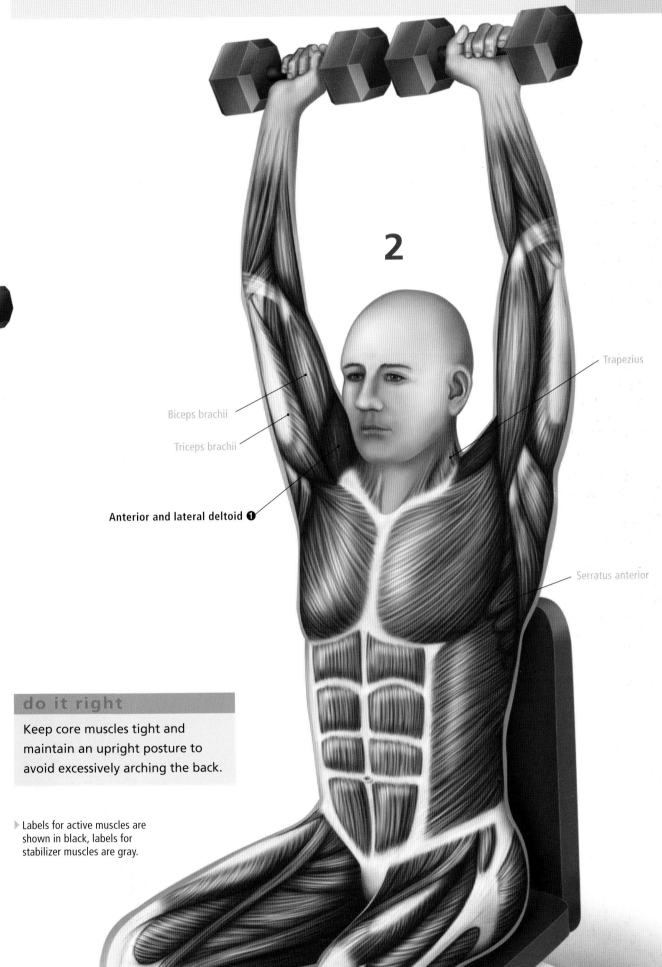

2

Biceps brachii

Triceps brachii

Anterior and lateral deltoid ❶

Trapezius

Serratus anterior

do it right

Keep core muscles tight and maintain an upright posture to avoid excessively arching the back.

▶ Labels for active muscles are shown in black, labels for stabilizer muscles are gray.

Front Raise

This exercise is great for improving strength in the front part of the shoulder. It targets the anterior deltoid muscles, while the rhomboids and trapezius stabilize the shoulder. The abdominals and the lower back muscles also need to work hard to maintain the upright posture, so they get a good secondary workout. The front raise is recommended for sports that require lifting and carrying, as well as combat sports such as martial arts and boxing, which rely on strength and endurance in the anterior deltoid to guard and strike. This exercise may also be used in shoulder rehabilitation, as it moves through a safe and controlled range of motion. Suitable for beginners and advanced trainers, the front raise can be performed at home or in a gym.

warning

Do not lift weights higher than shoulder level.

how to

Stand with feet shoulder-width apart and knees slightly bent. Grasp the barbell with hands shoulder-width apart, palms facing toward the body, and elbows straight or only slightly bent. Raise the barbell forward and upward until the weight is level with the shoulders. Slowly lower the weight down to the starting position.

variations

EASY

Grasp a dumbbell in each hand, palms facing toward the body with elbows straight or only slightly bent. Raise one dumbbell forward and upward until the weight is level with the shoulders. Slowly lower the weight down to the starting position, then repeat for the other side.

HARD

Perform the exercise one hand at a time using a low pulley cable. Stand with feet shoulder-width apart and knees slightly bent, facing away from the pulley. Contract the abdominal muscles tightly and raise the cable handle forward and upward until the weight is level with the shoulders. This variation puts a greater emphasis on the beginning of the lift.

active muscles

❶ Anterior deltoid
❷ Lateral deltoid
❸ Trapezius

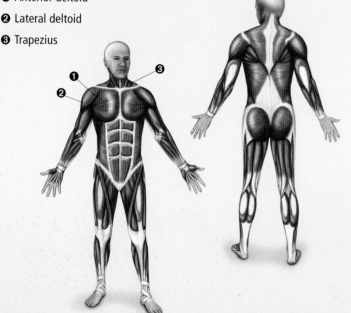

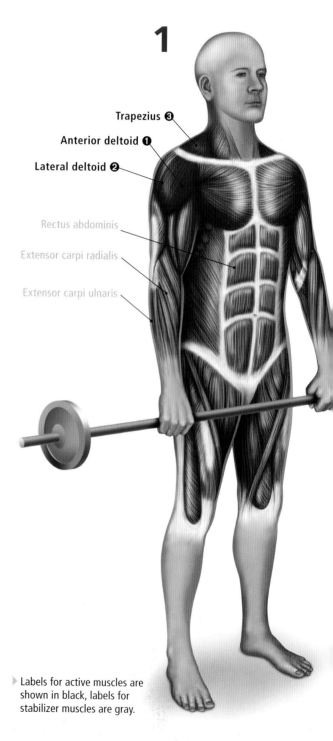

1

Trapezius ❸
Anterior deltoid ❶
Lateral deltoid ❷

Rectus abdominis
Extensor carpi radialis
Extensor carpi ulnaris

▶ Labels for active muscles are shown in black, labels for stabilizer muscles are gray.

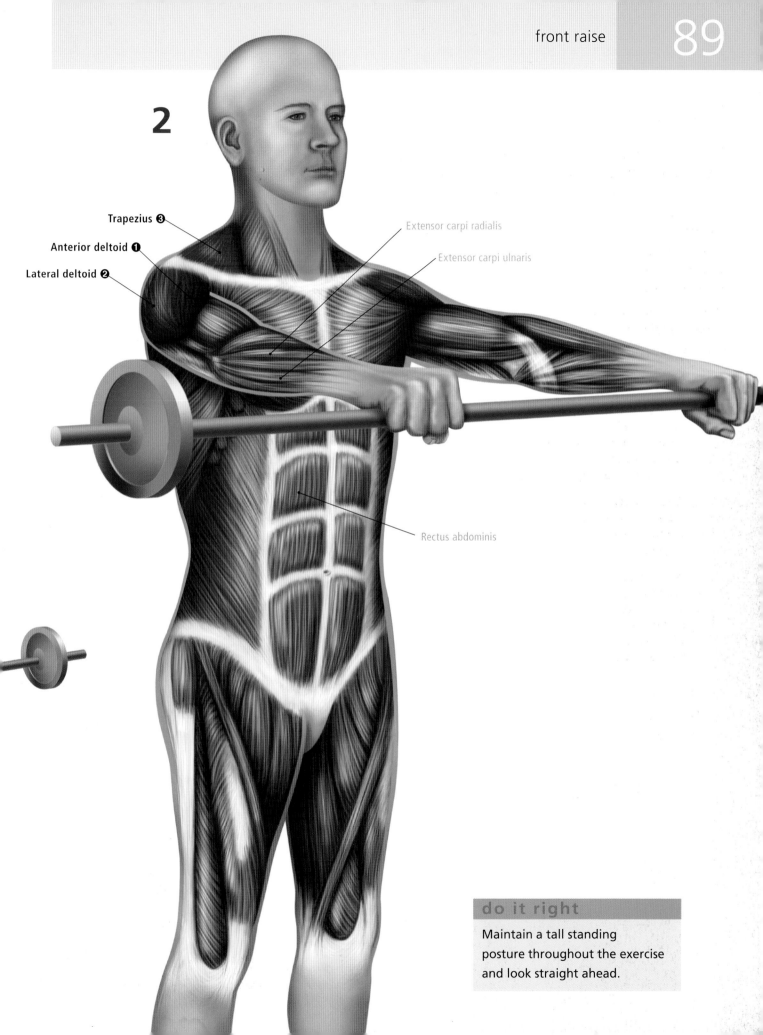

2

Trapezius ❸

Anterior deltoid ❶

Lateral deltoid ❷

Extensor carpi radialis

Extensor carpi ulnaris

Rectus abdominis

do it right

Maintain a tall standing posture throughout the exercise and look straight ahead.

Lateral Raise

This strength-building exercise is designed to specifically target the lateral aspect of the shoulder muscles. The trapezius and wrist extensor muscles also get a great workout, while the rotator cuff muscles are activated to stabilize the shoulder. It is suitable for beginners as well as more advanced trainers. The lateral deltoid is an important muscle for many sporting actions and increasing its size broadens the shoulders, which in effect narrows the appearance of the waist. This makes the lateral raise a perfect program inclusion for trainers who want to improve their body shapes. As it requires only a set of dumbbells, the lateral raise can be performed both at home and in the gym.

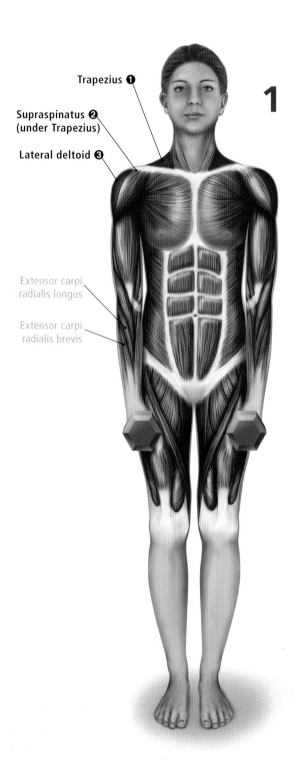

Extensor carpi radialis longus

Extensor carpi radialis brevis

Trapezius ❶

Supraspinatus ❷ (under Trapezius)

Lateral deltoid ❸

Extensor carpi radialis longus

Extensor carpi radialis brevis

1

how to

Stand with feet shoulder-width apart, knees slightly bent, and tilt forward at the hips slightly while maintaining a straight back. Hold a set of dumbbells in front of the thighs, with palms facing toward each other. Raise arms out to the side until elbows are at shoulder height. Make sure elbows remain slightly higher than wrists throughout the movement. Slowly lower the weights to the starting position.

variations

EASY

Train one side at a time if feeling uncomfortable lifting a dumbbell in each hand. Grasp a dumbbell in one hand and hold on to a solid stationary object for support with the other hand. Perform the exercise using the same technique and body position as outlined in the standard version.

HARD

To increase the level of difficulty at the start of the movement, use two low cable pulleys. Stand between two low pulleys; grasp the handle of the right pulley in the left hand, and the left pulley in the right hand. Raise arms out to the side until elbows are shoulder height. Make sure elbows remain slightly higher than wrists throughout the movement.

active muscles

❶ Trapezius

❷ Supraspinatus (under Trapezius)

❸ Lateral deltoid

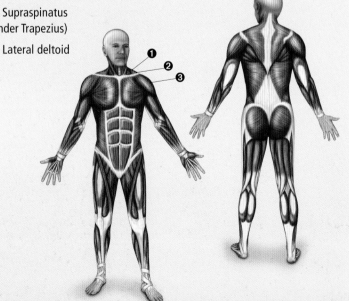

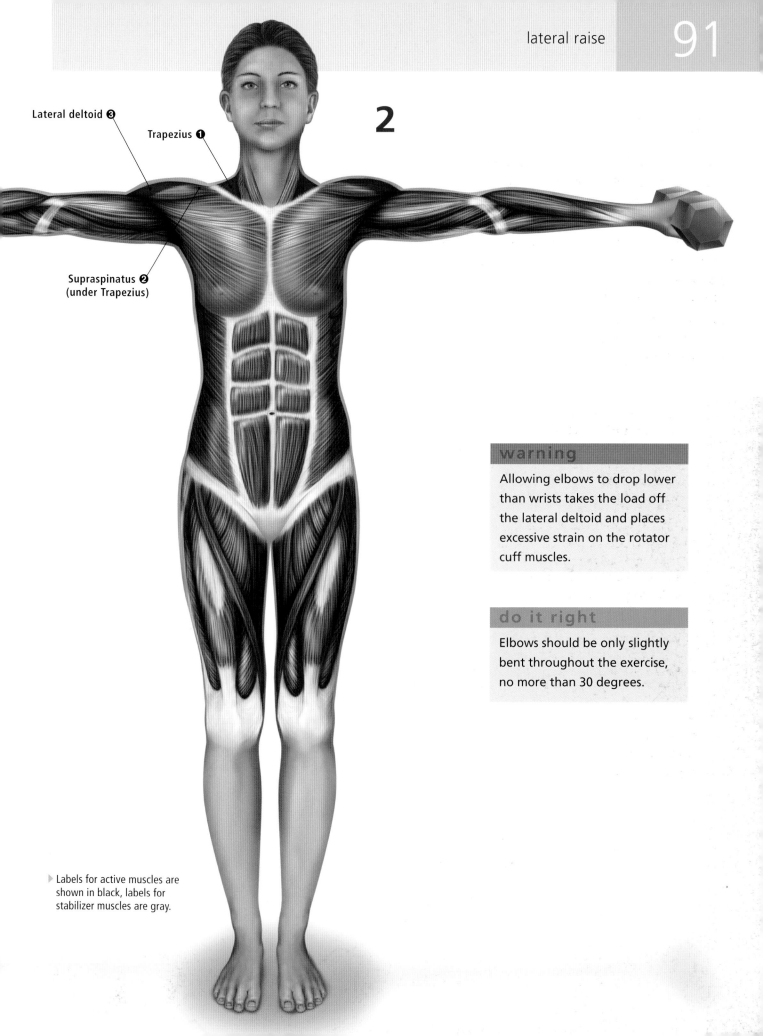

2

Lateral deltoid ❸

Trapezius ❶

Supraspinatus ❷
(under Trapezius)

warning

Allowing elbows to drop lower than wrists takes the load off the lateral deltoid and places excessive strain on the rotator cuff muscles.

do it right

Elbows should be only slightly bent throughout the exercise, no more than 30 degrees.

▷ Labels for active muscles are shown in black, labels for stabilizer muscles are gray.

Upright Row

This is an excellent exercise for improving shoulder strength, and targets the lateral deltoids. The anterior deltoid, trapezius, and biceps muscles are also called on to assist during the lift, while the muscles of the lower back must engage throughout the upright row to provide support to the upper body. This exercise is appropriate for beginners and advanced trainers, and is perfect for sports and occupations that involve lifting, dragging, or pulling movements. It is also a good starting point for trainers who want to learn more advanced, Olympic-style lifts, such as the jerk or snatch.

how to

Stand with feet shoulder-width apart and knees slightly bent. Grasp the barbell with hands shoulder-width apart, palms facing toward the body, and arms extended. Bend elbows and pull the barbell straight up, keeping the bar close to the body until the elbows are level with the shoulders. Allow the wrists to flex as the bar rises. Slowly lower the bar back to the starting position.

variations

EASY

Perform this exercise one side at a time using a dumbbell. Grasp a dumbbell in one hand and hold on to a solid stationary object for support with the other hand. Use the same technique and body position as the standard upright row, ensuring elbows point out to the side.

HARD

Use two dumbbells instead of a barbell to add extra instability, which requires greater control. Grasp the dumbbells with hands shoulder-width apart, palms facing toward the body, and arms extended. Bend elbows and pull the dumbbells straight up, keeping the weight close to the body until elbows are level with the shoulders. When the dumbbells are raised, wrists should be just below the shoulders with elbows pointed out to the sides.

active muscles

❶ Lateral deltoid
❷ Biceps brachii
❸ Brachioradialis
❹ Trapezius
❺ Supraspinatus (under Trapezius)

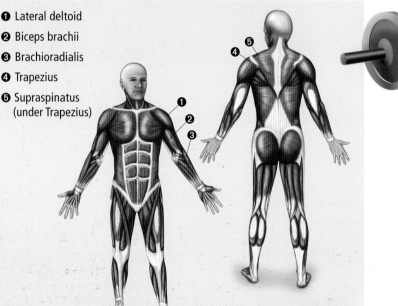

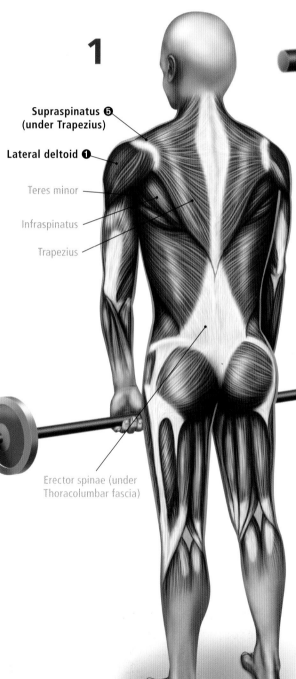

1

Supraspinatus ❺ (under Trapezius)

Lateral deltoid ❶

Teres minor

Infraspinatus

Trapezius

Erector spinae (under Thoracolumbar fascia)

2

Supraspinatus ❺
(under Trapezius)

Lateral deltoid ❶

Levator scapulae
(under Trapezius)

Teres minor

Infraspinatus

Trapezius

Erector spinae (under
Thoracolumbar fascia)

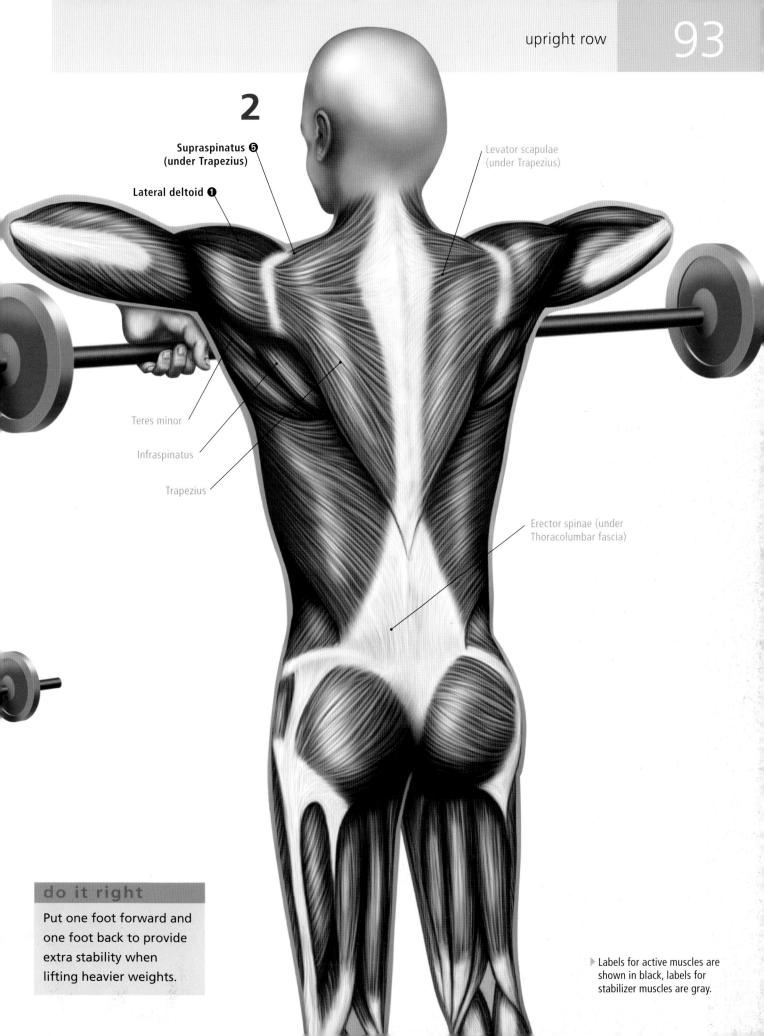

do it right

Put one foot forward and
one foot back to provide
extra stability when
lifting heavier weights.

▶ Labels for active muscles are
shown in black, labels for
stabilizer muscles are gray.

Shrug

This simple and effective strength-building exercise targets the back of the shoulders and the upper part of the back. The upper fibers of the trapezius muscle do most of the heavy lifting, with levator scapulae assisting. The shrug also gives the forearm muscles a great workout, as the relatively heavy load challenges the grip. This exercise is fantastic for improving strength in movements such as carrying, lifting, and dragging. It suits sports and occupations where these activities are required, and where grip strength is valued, such as rugby, wrestling, and martial arts. A barbell is all that is needed to perform the shrug. A sturdy rack to hold the bar at mid-thigh height will help to avoid injury while getting into the starting position.

how to

Stand with feet shoulder-width apart and knees straight. Grasp the barbell with hands shoulder-width apart, palms facing toward the body, and elbows straight. Allow the shoulders to sink downward. Pull the barbell straight up by elevating the shoulders toward the ears, keeping the bar close to the body. Slowly lower the bar back to the starting position.

variations

EASY

Use what is called a "mixed grip," where one palm faces toward the body and one faces away. This allows a stronger grip on the bar and stops the bar from rolling out of the hands. Pull the barbell straight up by elevating the shoulders toward the ears, keeping the bar close to the body. Slowly lower the bar back to the starting position.

HARD

Use one or two low cable pulleys, or two dumbbells, to provide an additional grip challenge and allow for a slightly greater and more natural range of motion. Grasp the dumbbell or pulley handles with hands to the side, below the shoulders, and with palms facing toward the body, elbows straight.

active muscles

❶ Trapezius

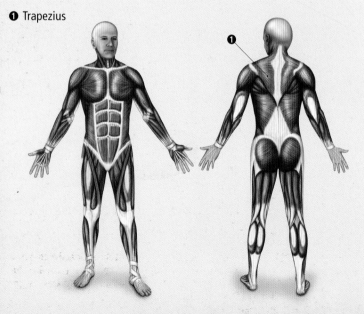

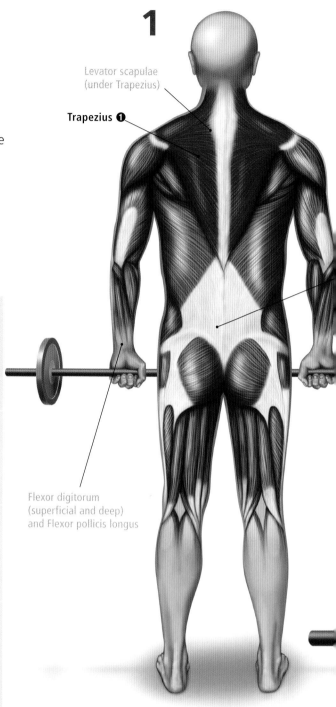

1

Levator scapulae (under Trapezius)

Trapezius ❶

Flexor digitorum (superficial and deep) and Flexor pollicis longus

warning

Ensure you use a rack to get the bar into the starting position, and maintain an upright posture during the exercise.

2

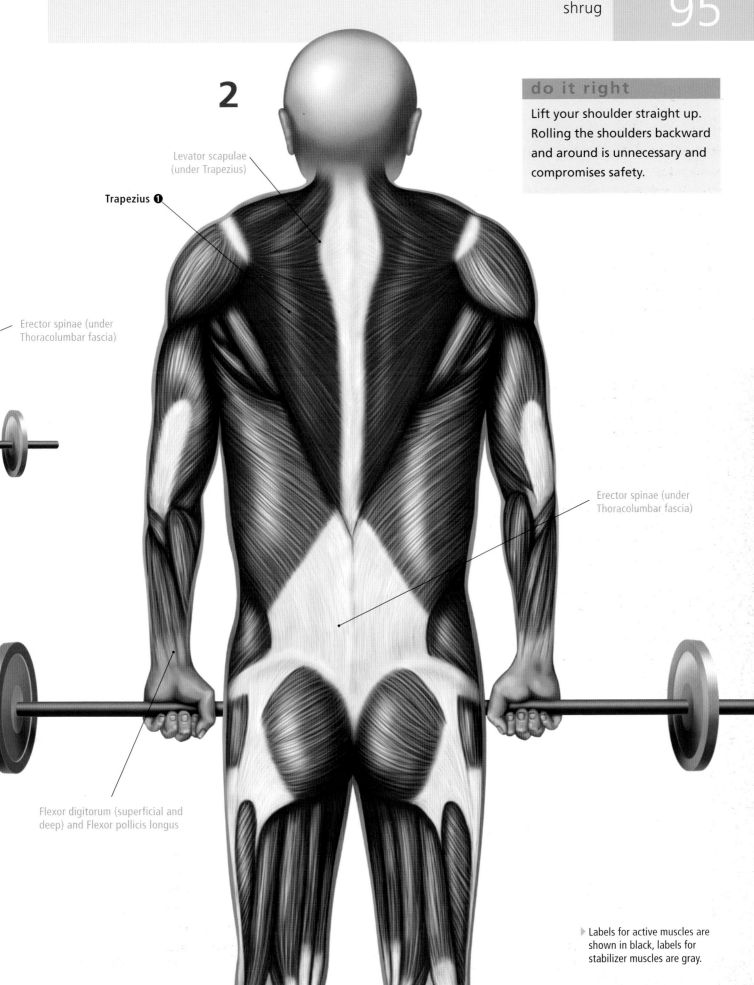

Levator scapulae
(under Trapezius)

Trapezius ❶

Erector spinae (under
Thoracolumbar fascia)

Erector spinae (under
Thoracolumbar fascia)

Flexor digitorum (superficial and
deep) and Flexor pollicis longus

▸ Labels for active muscles are
shown in black, labels for
stabilizer muscles are gray.

Wrist Curl

This simple exercise focuses on the wrist flexor muscles. The flexor carpi ulnaris and flexor carpi radialis are the main targets, while the muscles that provide grip strength also receive a great workout. Although not necessary as part of a beginner's training program, the wrist curl is a good addition at the end of a workout for more advanced trainers, to ensure their grip remains strong enough to perform other exercises as they start lifting heavier weights. Sports such as climbing, rugby, and martial arts, as well as all racket and bat sports, benefit from the high level of forearm and grip strength that develops with the wrist curl.

1

Flexor carpi ulnaris ❷

Flexor digitorum
(superficial and deep)

how to

Sit on a bench or chair and grasp a barbell with hands slightly narrower than shoulder-width apart, palms facing the ceiling. Rest forearms on the thighs, with wrists beyond the knees so that they can extend toward the floor without touching the legs. Lower the barbell toward the floor, allowing the bar to roll out of the palms and down into the fingers. Curl the barbell up by gripping the bar into the hand and pointing knuckles toward the ceiling.

variations

EASY

Stand up to perform the wrist curl if the wrist position in extension is uncomfortable. Hold the bar behind the body so that it is just below the gluteal muscles. Grasp the barbell with hands shoulder-width apart and palms facing away from the body. Pull the barbell up by gripping the bar up into the hands and curling the wrists.

HARD

For an extra challenge, perform the wrist curl using one dumbbell in each hand. This requires more control from other muscles in the forearm. Alternatively, perform the exercise with a barbell, palms facing the floor, to work the wrist extensor muscles and give the complete forearm a workout.

active muscles

❶ Flexor carpi radialis

❷ Flexor carpi ulnaris

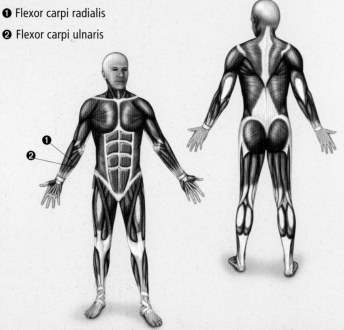

warning

Make sure the floor area is clear so that if the bar rolls out of the fingers it falls safely to the ground, without making contact with anyone's feet.

do it right

Keep wrists and elbows at the same height to maintain resistance on the wrist flexor muscles.

2

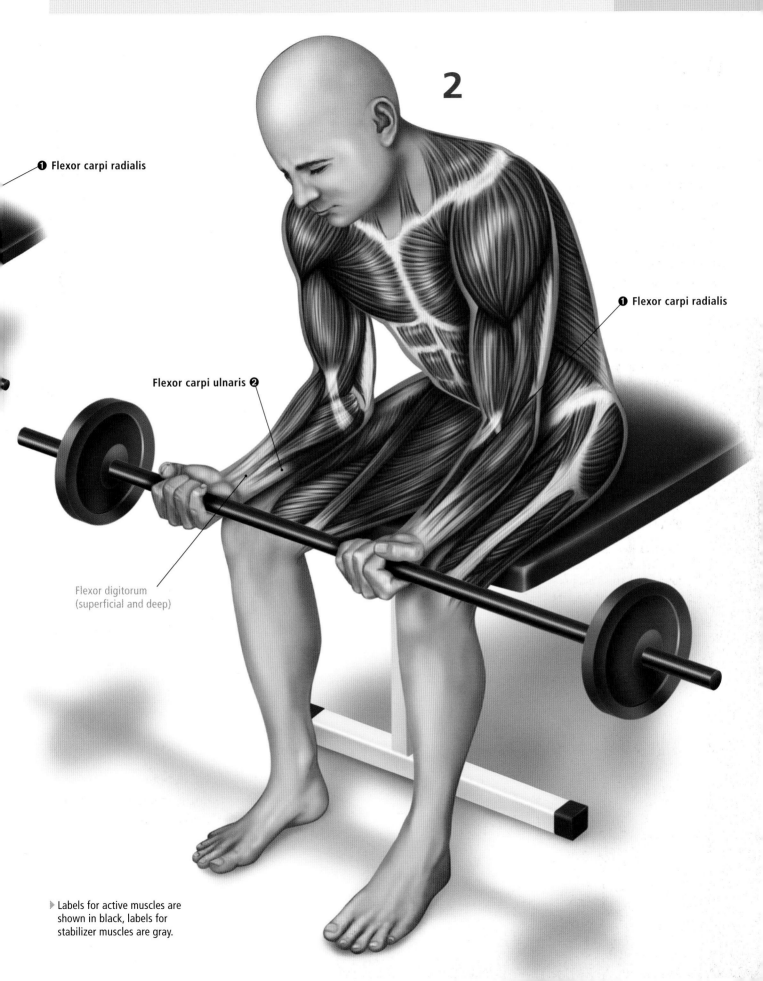

❶ Flexor carpi radialis

❶ Flexor carpi radialis

Flexor carpi ulnaris ❷

Flexor digitorum
(superficial and deep)

▷ Labels for active muscles are
shown in black, labels for
stabilizer muscles are gray.

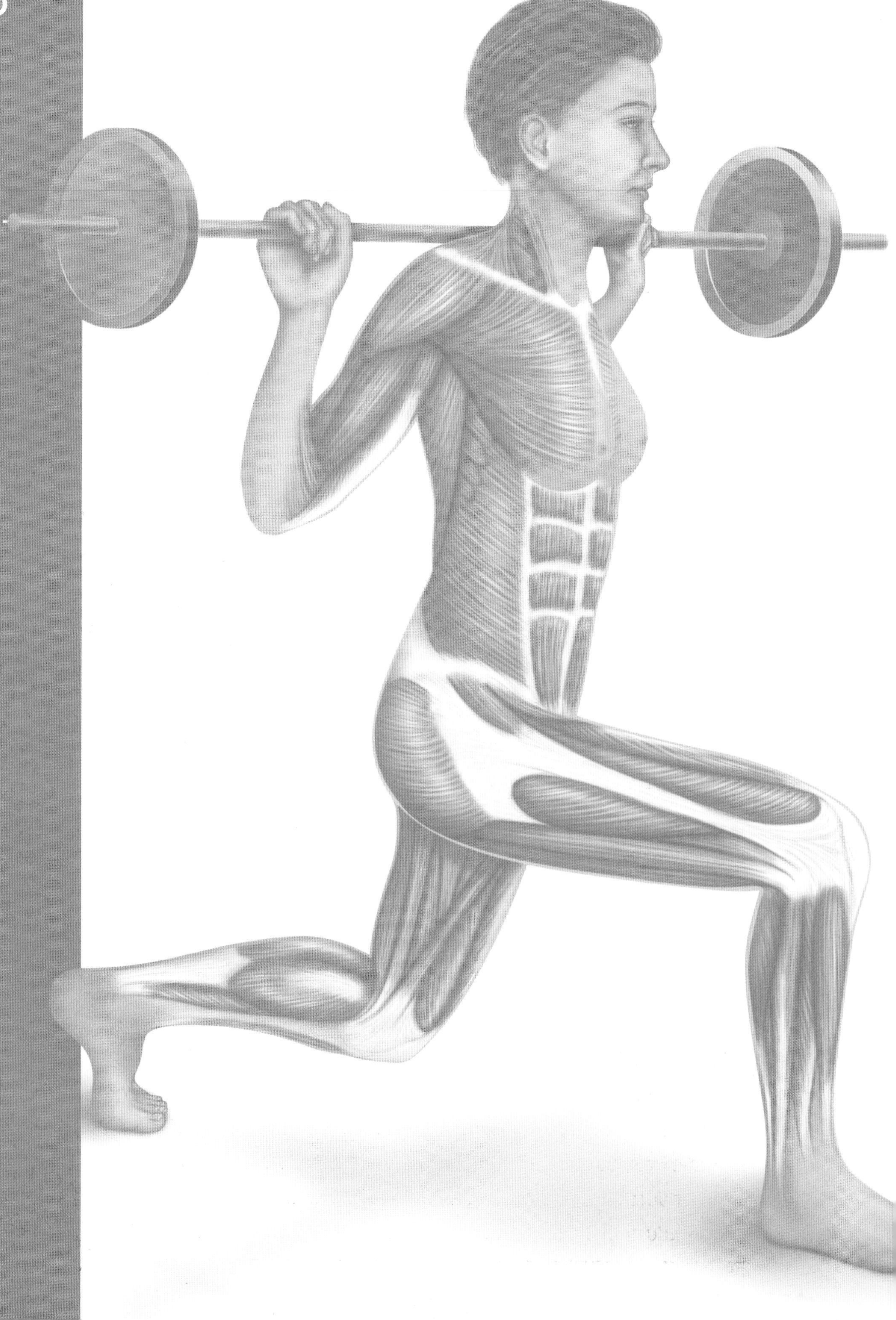

Leg and Buttock Exercises

This section focuses on the muscles required to walk, run, jump, and kick. Weak or injured leg and buttock muscles can cause considerable incapacity. Both beginners and experienced lifters should perform regular resistance training for these regions—to increase muscle strength, tone, power, endurance, and size. The exercises that follow are also suitable for rehabilitation programs, when allied health professionals prescribe training of specific intensity, volume, and frequency.

Do not neglect exercises that work the backs of the legs, as this can lead to hamstring-quadriceps strength imbalances. Target both sides of the leg to create a balanced workout.

Dumbbell Squat

This effective exercise works the quadriceps, the adductor muscle group, the buttocks, and to a lesser extent the hamstrings and lower back. It is excellent for any sports that require jumping, running, or kicking, and is often used by general fitness enthusiasts who want to firm the lower body. Although suitable for beginners, intermediate, and advanced exercisers, injuries can occur if the dumbbell squat is not performed correctly. Put the emphasis on correct technique. Beginners will feel more stable performing this exercise than the barbell squat, as the dumbbells are held close to the body's center of gravity. This exercise suits both home and gym environments.

how to

Stand with feet shoulder-width or a little wider apart, and hold dumbbells by the sides with palms facing each other. Focus forward, at about eye level. Bend knees and descend slowly, until the thighs are parallel to the floor. Return to the straight-leg position. Ensure that the back remains flat with a normal lumbar curve throughout the squat, and that heels stay on the floor. If heels start to lift, stop the descent.

variations

EASY

Use a stability ball against a wall, positioned at lower-back level, to provide extra support and balance during the squat. In the descent and ascent phases of the exercise, allow the ball to roll up and down the back, respectively.

HARD

Stand on the edge of a sturdy box or bench and perform a "single-leg dumbbell squat." Slowly lower the body as per the standard exercise, but drop one leg below the box and transfer full support of the bodyweight, plus dumbbells if appropriate, to the other leg. Adding external weight will really activate the quadriceps and gluteus maximus. Perform repetitions on each leg.

active muscles

❶ Adductor brevis (under Adductor longus)
❷ Vastus intermedius (under Rectus femoris)
❸ Adductor longus
❹ Adductor magnus
❺ Vastus lateralis
❻ Rectus femoris
❼ Vastus medialis
❽ Gluteus maximus

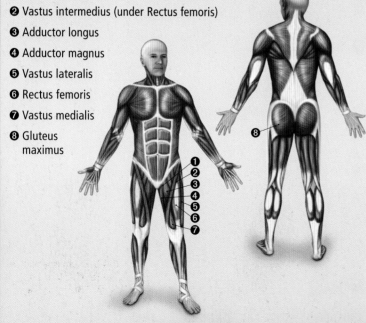

1

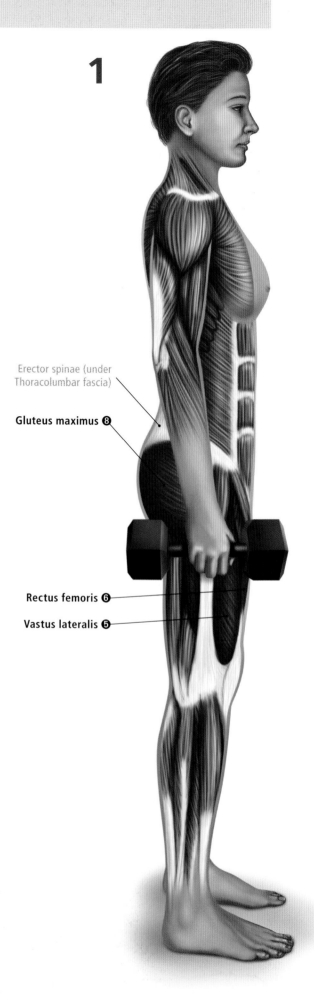

Erector spinae (under Thoracolumbar fascia)

Gluteus maximus ❽

Rectus femoris ❻

Vastus lateralis ❺

2

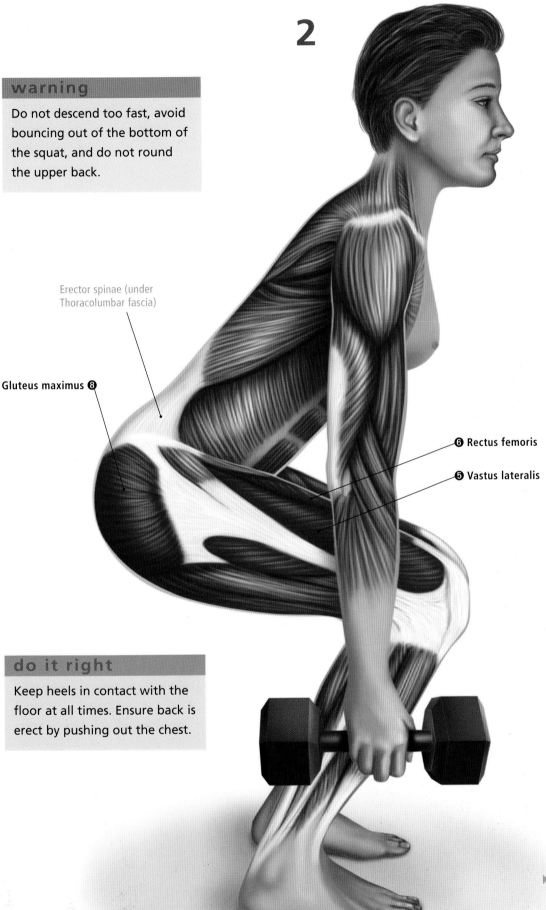

Erector spinae (under Thoracolumbar fascia)

Gluteus maximus **8**

6 Rectus femoris

5 Vastus lateralis

▶ Labels for active muscles are shown in black, labels for stabilizer muscles are gray.

Barbell Squat

This strengthening exercise targets the front of the thighs, the adductors, and the buttocks, as well as the hamstrings and the lower back to a minor extent. In particular, the quadriceps and the gluteus maximus receive a great workout. The barbell squat is ideal for sports that involve jumping, running, or kicking, and is a good lower body firming exercise for trainers. As the bar rests on the upper back during the movement, position it correctly to safeguard against injury. Ensure that proper technique is maintained throughout the exercise before increasing the weight. The barbell squat requires minimal equipment, so it can be performed either at home or in the gym.

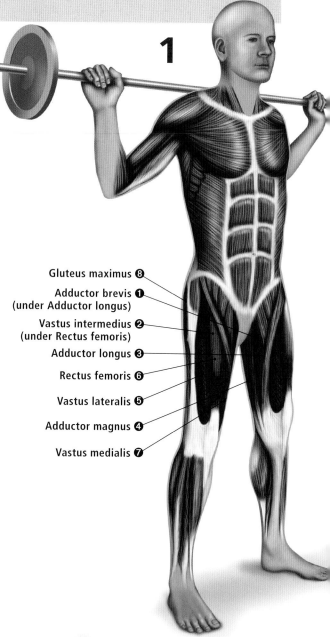

Gluteus maximus ❽
Adductor brevis ❶
(under Adductor longus)
Vastus intermedius ❷
(under Rectus femoris)
Adductor longus ❸
Rectus femoris ❻
Vastus lateralis ❺
Adductor magnus ❹
Vastus medialis ❼

how to

Stand with feet shoulder-width apart, place the barbell across the shoulders, and rest the bar on the upper level of the trapezius muscle. Grip the bar slightly wider than shoulder-width apart. Focus forward, at about eye level—do not look down. Bend knees and descend slowly, keeping heels on the floor. Stop when thighs are parallel to the floor, then return to the straight-leg position. If heels do start to lift, stop the descent at that point.

variations

EASY

Try a "quarter squat," which uses the same technique as the standard exercise, but limits the range of motion. Stop the descent when thighs are halfway between being upright and being parallel to the floor, then return to the starting position.

HARD

Try a "front squat." Before starting, significantly reduce the amount of weight being lifted. Rest the barbell across the front of the shoulders, with palms facing toward the body and fingertips just gripping the bar. This grip extends the wrist muscles. Follow the standard exercise, but keep the torso in a more upright position to avoid the barbell slipping off the shoulders. Ensure elbows face forward and upper arms stay parallel to the ground.

active muscles

❶ Adductor brevis
(under Adductor longus)
❷ Vastus intermedius
(under Rectus femoris)
❸ Adductor longus
❹ Adductor magnus
❺ Vastus lateralis
❻ Rectus femoris
❼ Vastus medialis
❽ Gluteus maximus

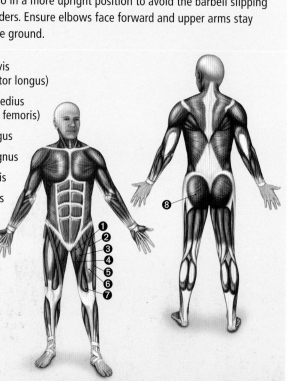

do it right

Keep the back flat, with its normal lumbar curve, throughout the movement. Avoid rounding back forward during descent or ascent.

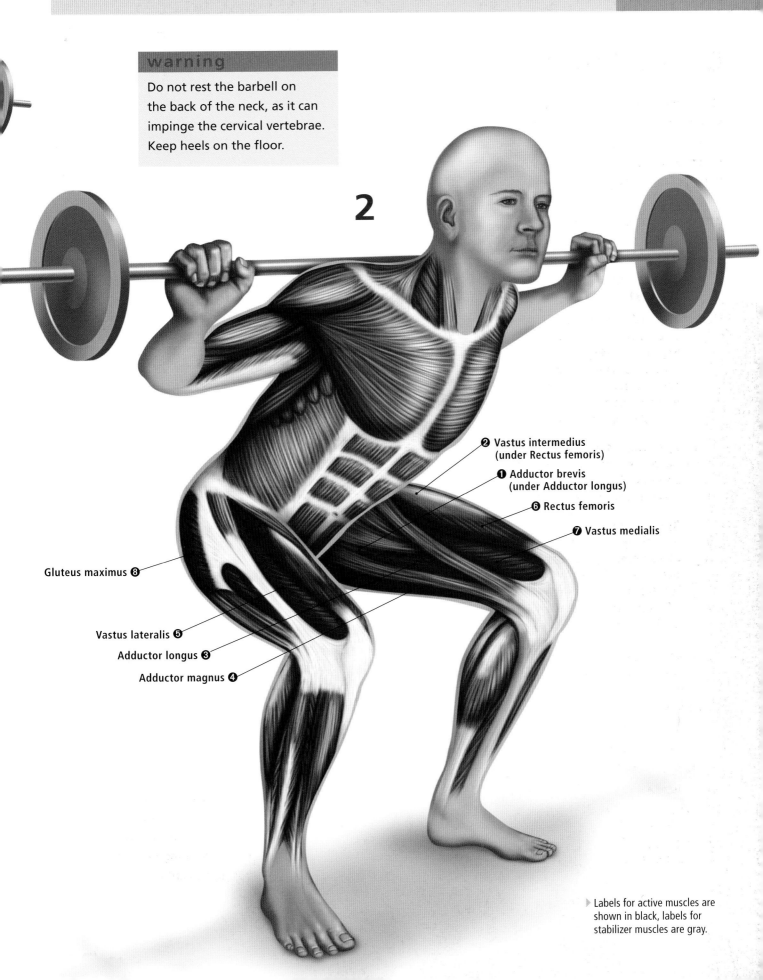

warning

Do not rest the barbell on the back of the neck, as it can impinge the cervical vertebrae. Keep heels on the floor.

2

❷ Vastus intermedius
(under Rectus femoris)

❶ Adductor brevis
(under Adductor longus)

❻ Rectus femoris

❼ Vastus medialis

Gluteus maximus ❽

Vastus lateralis ❺

Adductor longus ❸

Adductor magnus ❹

▷ Labels for active muscles are shown in black, labels for stabilizer muscles are gray.

Barbell Lunge

This popular exercise works the quadriceps, buttocks, and adductor muscle groups, which include the vastus lateralis, vastus intermedius, vastus medialis, gluteus maximus, adductor brevis, and adductor magnus muscles. It is great for sports that encompass running, jumping, and kicking, as the lunge involves movement at the hip, knee, and ankle joints. Although lighter weights are used in this exercise compared to the various squats, the advantage of the lunge is that it involves stepping out with a single leg—a common movement in many sports. It is suitable for both beginner and advanced exercisers, and can be performed with a barbell, dumbbells, or simply using bodyweight in a home or gym environment.

how to

Stand with feet together and hold barbell across the upper trapezius, with hands slightly wider than shoulder-width apart. Take a large step forward, placing the front foot on the ground with toes facing forward. Bend first the front knee, then both knees as the body lowers. Stop when the back leg is close to a 90-degree angle, with heel raised and weight on the ball of the foot. Push up to the starting position using the front leg.

variations

EASY

Hold a dumbbell in each hand by the sides, rather than a barbell across the back if finding it hard to stay balanced. This increases stability throughout the movement because the weights are closer to the body's center of gravity.

HARD

Perform a "walking lunge." Follow the standard exercise, but at the end of each lunge step forward with the back leg so that it assumes the forward lunge position. Alternatively, try using dumbbells by the sides, or holding a weight plate or medicine ball above the head with arms extended.

active muscles

❶ Adductor brevis (under Adductor longus)
❷ Vastus intermedius (under Rectus femoris)
❸ Adductor magnus
❹ Vastus lateralis
❺ Vastus medialis
❻ Gluteus maximus

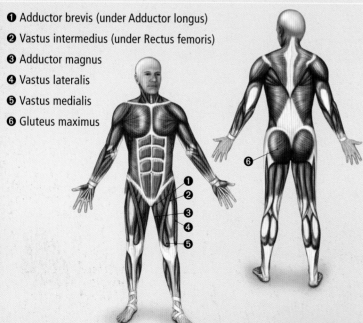

1

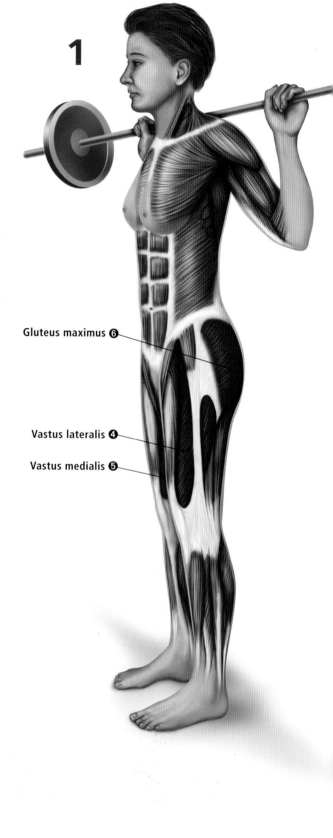

Gluteus maximus ❻

Vastus lateralis ❹

Vastus medialis ❺

2

Look straight ahead throughout the movement and maintain an erect torso.

Vastus lateralis ❹

❻ Gluteus maximus

❸ Adductor magnus

Vastus lateralis ❹

Vastus medialis ❺

Do not descend too rapidly. Avoid using heavy weights in the "walking lunge" variation of this exercise.

▶ Labels for active muscles are shown in black, labels for stabilizer muscles are gray.

Deadlift

This structural exercise works multiple muscles, especially those in the buttocks, legs, and back. When performed with the correct technique, the deadlift is suitable for beginners through to advanced lifters. And because it requires only a barbell and weight plates, it is suitable for both home and gym environments. Those new to resistance training should start with light weights and increase weight slowly, so their bodies can adapt to the increased loading and they can perfect their technique. Advanced lifters can perform the deadlift with heavy weights.

1

Erector spinae (under Thoracolumbar fascia)

❼ Gluteus maximus

Semimembranosus and Semitendinosus

❹ Vastus lateralis

Biceps femoris

how to

Stand in front of a weighted barbell placed on the ground, with feet shoulder-width apart. Keeping back and arms straight, bend at the knees until able to grasp the bar using an alternating grip, with hands just outside the knees. Look forward, brace the abdominal and lower back muscles, then straighten knees and hips until fully extended. Shrug shoulders toward the ears at the top of the lift. Lower shoulders, bend knees, and slowly return the barbell to the floor.

variations

EASY

Use a shortened bar, attached to the cable of a pin-loaded weight stack. This makes adjusting the weight easier, and because the weight stack is guided on metal columns, it provides greater stability to the moving bar.

HARD

Increase the challenge by performing a "sumo deadlift." Adopt a wide stance with knees bent, feet pointing slightly outward, and eyes focused directly ahead. Grip the barbell on the inside of the legs, with hands close together in an alternating grip, and arms straight. Keep the back straight, brace abdominal and lower back muscles, then extend the legs. Slowly bend the knees to return the bar to the floor.

active muscles

❶ Adductor brevis (under Adductor longus)
❷ Vastus intermedius (under Rectus femoris)
❸ Adductor magnus
❹ Vastus lateralis
❺ Rectus femoris
❻ Vastus medialis
❼ Gluteus maximus

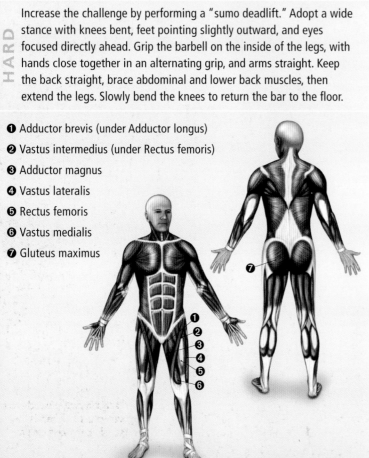

do it right

Maintain a flat back with normal lumbar curve. Keep arms extended throughout the movement and lower weight slowly to the floor.

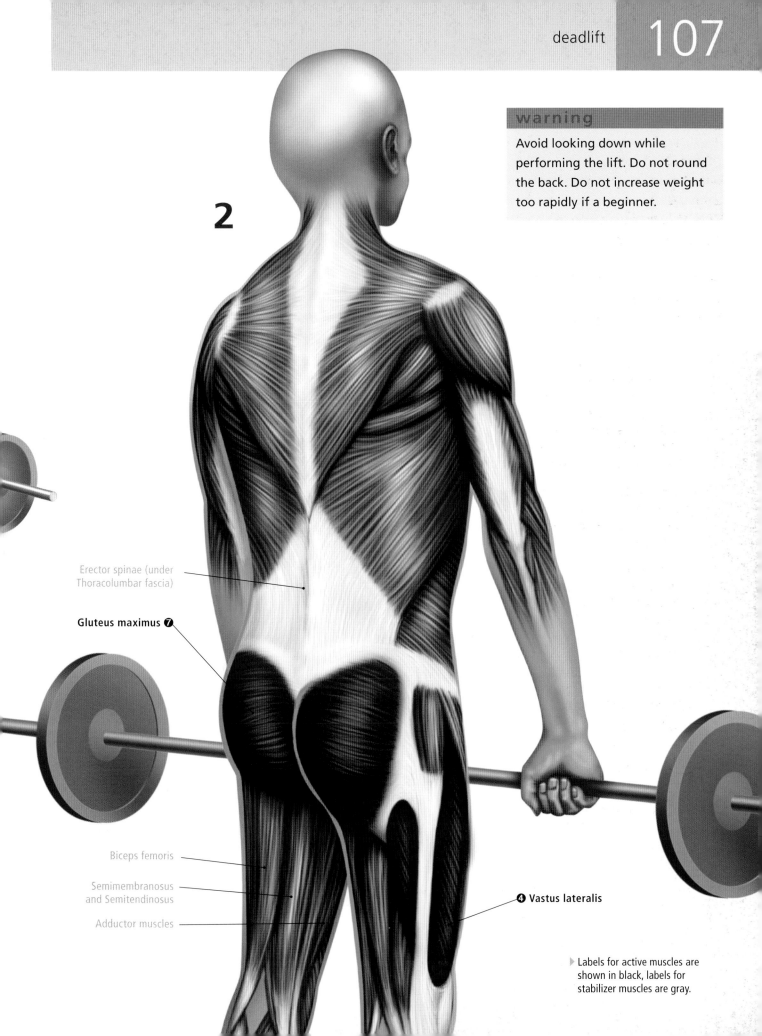

2

warning

Avoid looking down while performing the lift. Do not round the back. Do not increase weight too rapidly if a beginner.

Erector spinae (under Thoracolumbar fascia)

Gluteus maximus ❼

Biceps femoris

Semimembranosus and Semitendinosus

Adductor muscles

❹ Vastus lateralis

▶ Labels for active muscles are shown in black, labels for stabilizer muscles are gray.

Romanian Deadlift

This variation of the deadlift is named after the famous Romanian weightlifter, Nicu Vlad, who employed this exercise while training in the United States. It targets the buttocks and hamstring muscle groups, in particular working the gluteus maximus, semitendinosus, semimembranosus, and biceps femoris. Although all levels of lifter can perform the Romanian deadlift, make sure the correct technique is used to avoid injury. In this type of deadlift, there is minimal bending of the knees, so lift less weight in comparison to what is used in other deadlift variations. The Romanian deadlift is suitable for both the home or gym environment, as it requires only a barbell or set of dumbbells to execute.

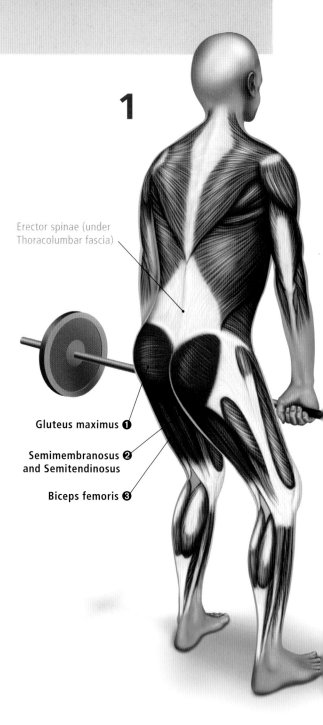

1

Erector spinae (under Thoracolumbar fascia)

Gluteus maximus ❶

Semimembranosus ❷ and Semitendinosus

Biceps femoris ❸

how to

Stand with feet shoulder-width apart and grasp a barbell with hands just outside the thighs, palms facing toward the body. Bend the knees slightly and retain this knee angle throughout the movement. Keeping the bar close to the body and the back and arms straight, slowly lower the trunk by bending at the hips until the bar reaches a position equal to, or slightly above, the knees. Extend the hips and return to the starting position.

variations

EASY
Use bodyweight only, rather than a barbell, and focus on technique. Once confident that technique is sound, progress to a barbell or try dumbbells instead.

HARD
To increase the challenge, perform a "single-leg Romanian deadlift." Hold a dumbbell in each hand at hip level, close to the body, with knees slightly bent. Bend forward at the hips as per the standard exercise, but allow one leg to extend backward as the trunk descends. Keep the back straight. Lift the leg until it forms a flat plane with the back. Slowly return to the starting position, then repeat on the other leg.

active muscles

❶ Gluteus maximus

❷ Semimembranosus and Semitendinosus

❸ Biceps femoris

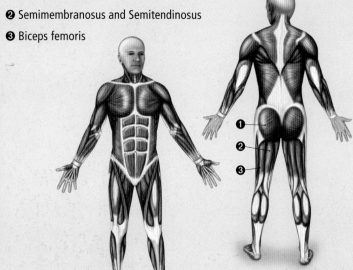

2

Erector spinae (under Thoracolumbar fascia)

Gluteus maximus ❶

Semimembranosus ❷
and Semitendinosus

Biceps femoris ❸

warning

Take care not to use excessive
weight. Do not round the back.

▶ Labels for active muscles are
shown in black, labels for
stabilizer muscles are gray.

Step-up

This classic exercise focuses on many muscles of the lower body, such as the quadriceps, gluteus maximus, hip flexors (iliopsoas), calves, and to a minor extent the hamstrings. It is therefore a great exercise for sports such as football, hockey, and lacrosse. With a little practice, the step-up is easy to perform. It can be incorporated into most exercise routines and is a popular choice for circuit training programs. The step-up is suitable for all levels of exercisers. It requires just a barbell or pair of dumbbells, and a sturdy box or step, which makes it easy to perform both at home and in the gym.

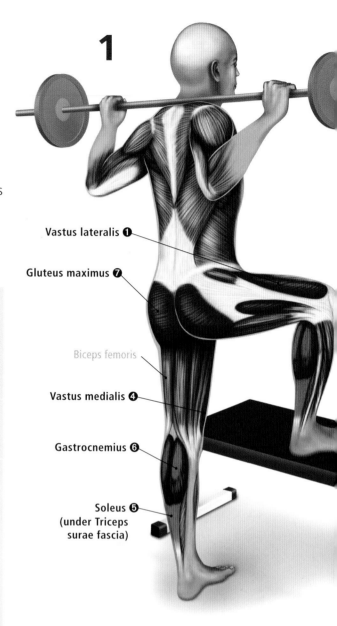

1

Vastus lateralis **❶**

Gluteus maximus **❼**

Biceps femoris

Vastus medialis **❹**

Gastrocnemius **❻**

Soleus **❺**
(under Triceps
surae fascia)

how to

Stand in front of a sturdy box, ranging from 8–16 inches (20–40 cm) in height, with feet close together. Place a weighted barbell on the upper trapezius muscles—avoid contact with the neck. Looking forward and keeping a straight back, raise one leg off the ground and place foot on the box. Step up, placing the other foot on the box. Step down one leg at a time in a controlled movement.

variations

EASY

Hold dumbbells by the sides with palms facing toward the body. This keeps the weight closer to the body's center of gravity and is therefore less challenging from a balance perspective.

HARD

Step up on the box with one foot, but instead of placing the other foot on the box, continue lifting that leg until the thigh is parallel to the ground before placing it down on the box or directly back on the floor. This additional movement requires greater hip flexion and increases the balance challenge.

active muscles

❶ Vastus lateralis

❷ Vastus intermedius
(under Rectus femoris)

❸ Rectus femoris

❹ Vastus medialis

❺ Soleus (under
Triceps surae fascia)

❻ Gastrocnemius

❼ Gluteus maximus

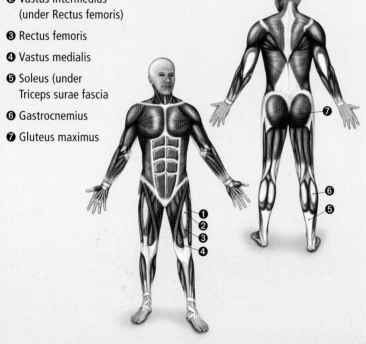

warning

Take care when stepping down from the box. Ensure barbell is off the back of the neck at all times.

2

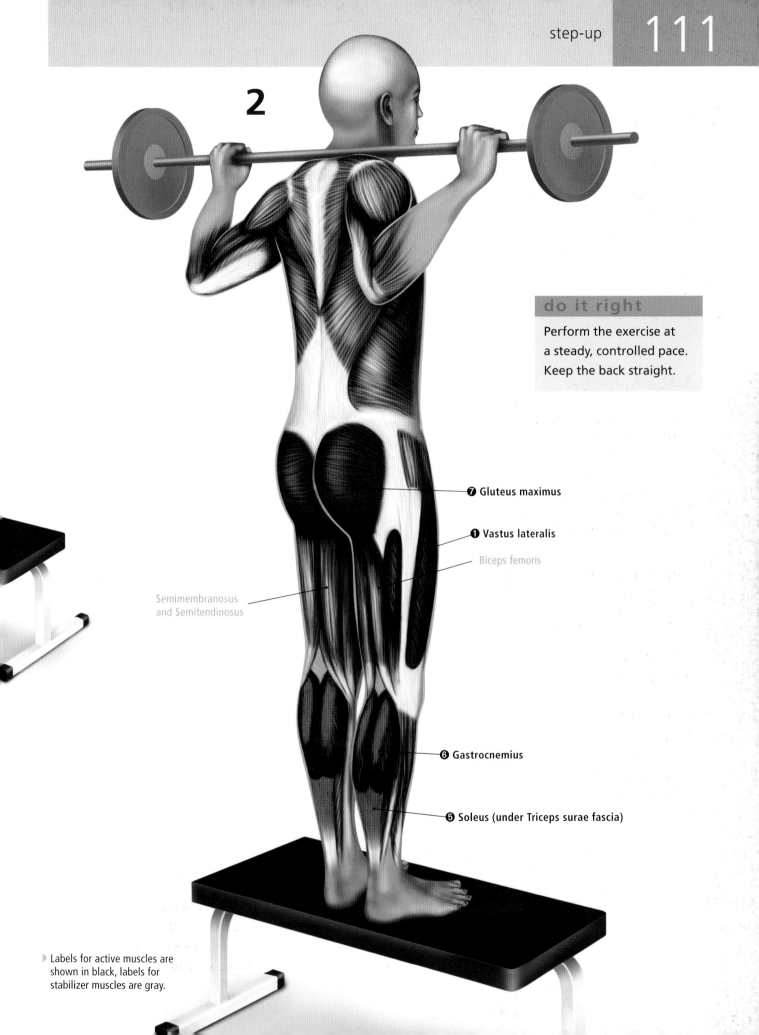

Perform the exercise at a steady, controlled pace. Keep the back straight.

❼ **Gluteus maximus**

❶ **Vastus lateralis**

Biceps femoris

Semimembranosus and Semitendinosus

❻ **Gastrocnemius**

❺ **Soleus (under Triceps surae fascia)**

▷ Labels for active muscles are shown in black, labels for stabilizer muscles are gray.

Standing Calf Raise

This leg exercise trains the plantar flexor muscles of the calf, which are used to point the foot in actions such as pushing the accelerator pedal of a car. It is a good exercise choice for beginners through to advanced lifters, and can be performed in a home or gym environment, depending on the variation used and equipment that is available. Experienced lifters often use heavy weights for the standing calf raise. However, beginners should start with light weights and progress only when their technique and strength improve. This exercise benefits sports that involve running or jumping, such as football, hockey, track sprinting, and gymnastics. It is also useful for rehabilitation training for the ankles or calves.

how to

Load the desired weight onto the standing calf-raise machine. Place shoulders on the padded supports and feet on the footplate, with heels slightly protruding from the end of the plate. Keeping a straight back, stand until the legs are extended to assume the starting position. Rise as high as possible on the balls of the feet, then slowly lower the heels to the lowest position possible below the footplate. Do not bend at the knees or hips during the movement.

variations

EASY

Stand with both feet on a step or sturdy box and follow the standard exercise. As strength builds, progress to standing on one leg, then to holding a dumbbell, both of which increase the stress on the calf muscles. This is a good option for home training, when a standing calf-raise machine may not be available.

HARD

Ramp up the workout with eccentric overload training, one leg at a time. Perform the upward movement with both calves, lifting the weight against gravity, but slowly resist the downward movement with only one calf, relaxing the opposite calf. Repeat until the desired number of downward movements has been completed with each calf.

active muscles

❶ Gastrocnemius
❷ Soleus (under Triceps surae fascia)
❸ Flexor hallucis longus
❹ Flexor digitorum longus
❺ Tibialis posterior
❻ Plantaris

(3, 4 and 5 all under Soleus and Gastrocnemius)

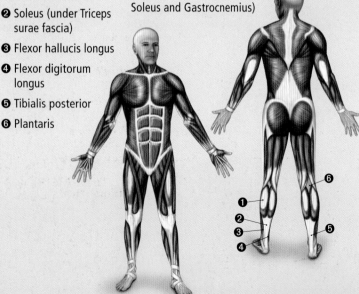

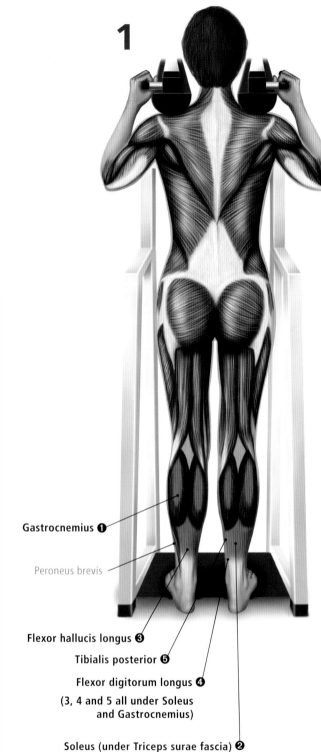

1

Gastrocnemius ❶

Peroneus brevis

Flexor hallucis longus ❸

Tibialis posterior ❺

Flexor digitorum longus ❹

(3, 4 and 5 all under Soleus and Gastrocnemius)

Soleus (under Triceps surae fascia) ❷

warning

Do not bounce at the bottom of the movement. Stay slow and controlled as heels descend.

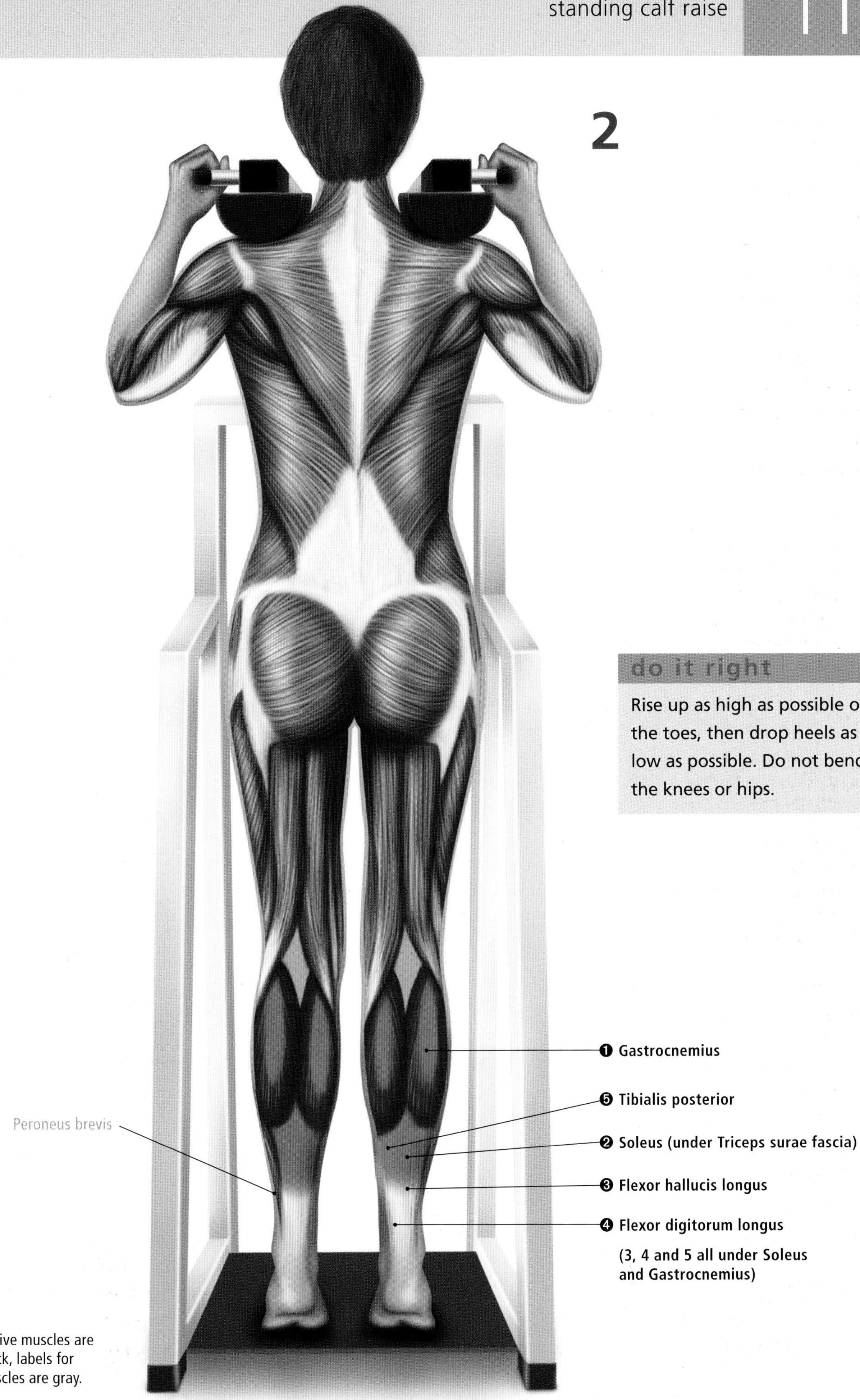

2

do it right

Rise up as high as possible on the toes, then drop heels as low as possible. Do not bend the knees or hips.

Peroneus brevis

❶ Gastrocnemius

❺ Tibialis posterior

❷ Soleus (under Triceps surae fascia)

❸ Flexor hallucis longus

❹ Flexor digitorum longus

(3, 4 and 5 all under Soleus and Gastrocnemius)

▷ Labels for active muscles are shown in black, labels for stabilizer muscles are gray.

Seated Calf Raise

This calf exercise is another good choice for training the plantar flexor muscles. The bent-knee position that it requires minimizes the involvement of the large gastrocnemius muscle and places more emphasis on the other plantar flexor muscles of the calf. Use it in training for sports such as track sprinting, football, gymnastics, and hockey. This exercise can also be successfully included in rehabilitation programs for the calves and ankles. The seated calf raise is appropriate for all levels of trainer and it is possible to perform in a home or gym environment. A specialized machine is generally used for the standard exercise; however, other variations that do not require a machine are just as effective.

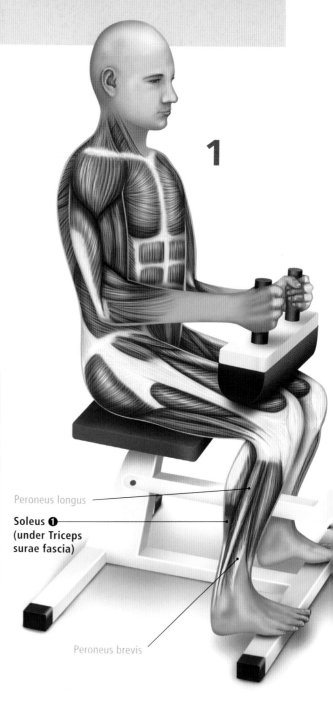

Peroneus longus

Soleus ❶
(under Triceps
surae fascia)

Peroneus brevis

how to

Load the desired weight onto the seated calf-raise machine. Sit down and place lower thighs and knees under the padded support, and feet on the footplate, with heels protruding from the back of the plate. Lift the padded support by rising onto the balls of the feet. Slowly lower the weight until heels drop as far below the footplate as possible. Repeat sets as desired.

variations

EASY

Sit on a chair or bench and place feet on a low box or step. Place a folded towel across the lower thighs and knees, then rest a loaded barbell on the towel. Lift the heels until balancing on the balls of the feet. Slowly lower the heels as far as possible below the box.

HARD

Increase the challenge by focusing on one calf at a time. Perform the upward movement as per the standard exercise, with both calves lifting the weight against gravity. Slowly resist the downward movement with only one calf, relaxing the opposite calf. This applies eccentric overload to the muscles in the lowering phase of the exercise. Repeat with the alternate leg.

active muscles

❶ Soleus (under Triceps surae fascia)
❷ Flexor hallucis longus
❸ Flexor digitorum longus
❹ Tibialis posterior
❺ Plantaris

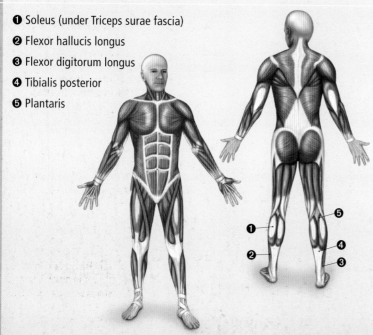

do it right

Rise as high as possible on the balls of the feet during the upward movement, and drop heels as low as possible during the downward movement.

2

warning

Maintain a slow, controlled technique and do not bounce at the bottom of the movement.

Peroneus longus

❶ **Soleus (under Triceps surae fascia)**

Peroneus brevis

▶ Labels for active muscles are shown in black, labels for stabilizer muscles are gray.

Leg Extension

This popular exercise targets the quadriceps muscle group. It is useful for any sports that involve running, kicking, jumping, or skipping movements. The leg extension is suitable for beginners through to advanced exercisers, and can be executed in a home or gym environment, depending on which variation is used. As it is generally performed seated on a specialized machine, this exercise does not require the same level of balance as squat and lunge exercises. Although this can be an advantage in some circumstances, such as early rehabilitation scenarios where balance may be problematic, it does make the leg extension less useful for trainers who are looking for balance and stability challenges in their routines.

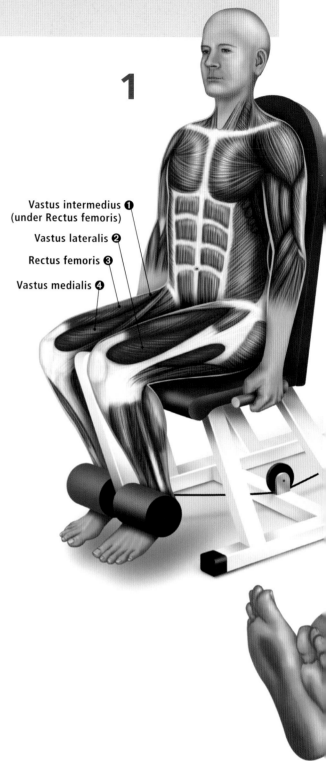

1

Vastus intermedius ❶
(under Rectus femoris)

Vastus lateralis ❷

Rectus femoris ❸

Vastus medialis ❹

how to

Sit on a leg-extension machine and adjust the back rest so the back of the knees rest against the edge of the padded seat and knee joints align with the rotation axis of the lever arm. Adjust the leg pad to just above the front of the ankles. Pushing against the leg pad, lift legs until fully extended. Keep the torso upright and feet pointing forward. Slowly return to the starting position, then repeat.

variations

EASY

Use ankle weights and perform the exercise sitting on a chair or sturdy table. Position the weight cuff directly above the ankle of each leg. If using a chair, ensure that it is tall enough to avoid feet touching the ground on the return phase of the movement. Keep torso straight throughout the exercise.

HARD

Go for the extra challenge of eccentric overload when lowering weight, one leg at a time. Perform the standard exercise, using both legs to raise the weight until legs are fully extended. However, on the return phase of the movement, resist the pull of the weight with only one leg. Complete repetitions on each leg.

active muscles

❶ Vastus intermedius (under Rectus femoris)

❷ Vastus lateralis

❸ Rectus femoris

❹ Vastus medialis

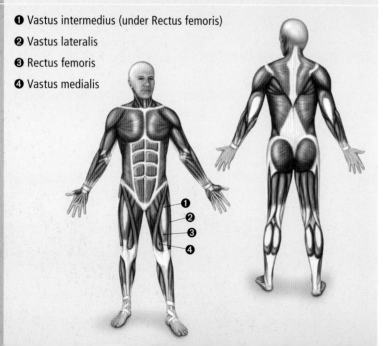

warning

Do not descend too rapidly or swing the trunk.

2

do it right

Extend the legs until they are straight. Descend slowly and maintain an erect torso throughout the exercise.

Vastus intermedius ❶
(under Rectus femoris)

Vastus lateralis ❷

Rectus femoris ❸

Vastus medialis ❹

▶ Labels for active muscles are shown in black, labels for stabilizer muscles are gray.

Seated Leg Curl

This seated exercise is another popular choice for homing in on the semitendinosus, semimembranosus, and biceps femoris muscles of the hamstrings. Like the lying leg curl, it provides great training for sports that involve running and kicking movements, such as track, field, and football. The seated leg curl is suitable for beginners and advanced exercisers; however, because it requires equipment to execute, it is not usually performed at home. Many trainers prefer this exercise to the lying leg curl because they find the seated position more comfortable.

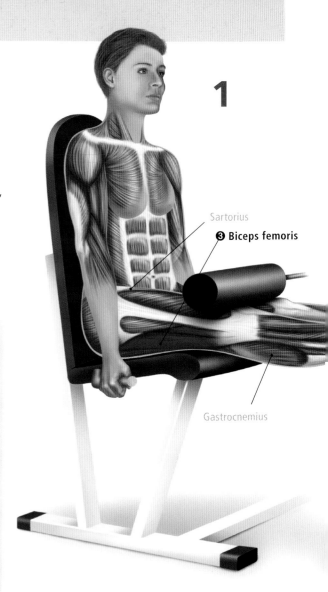

1

Sartorius

❸ Biceps femoris

Gastrocnemius

how to

Sit in a seated leg-curl machine and adjust the back support so that knees rest against the edge of the padded seat, with knee joints aligned with the rotation axis of the lever arm. Level leg pads with the back of the ankle. Hold the handgrips and pull against the leg pads, curling legs as far as the machine will allow toward the buttocks. Slowly return to the starting position and repeat. Keep torso upright and feet pointing forward throughout the exercise.

variations

EASY

Before beginning, reduce the amount of weight to the lightest plate possible, then follow the standard exercise.

HARD

Follow the standard exercise, curling both legs toward the buttocks. However, on the return phase, resist the pull of the weight with only one leg. This places greater stress on the hamstring muscles during the lengthening phase, as this motion (eccentric phase) is capable of greater force than the shortening phase (concentric phase) when the legs move toward the buttocks. Alternate legs until the desired number of repetitions have been completed.

active muscles

❶ Semitendinosus
❷ Semimembranosus
❸ Biceps femoris

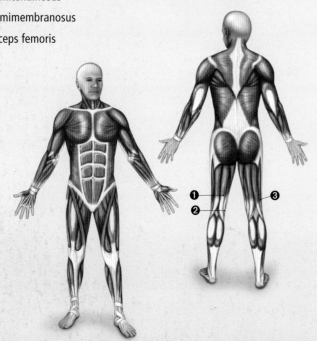

do it right

Curl the legs toward the buttocks as far as the leg pad will allow. Return slowly to the starting position and maintain an erect torso.

2

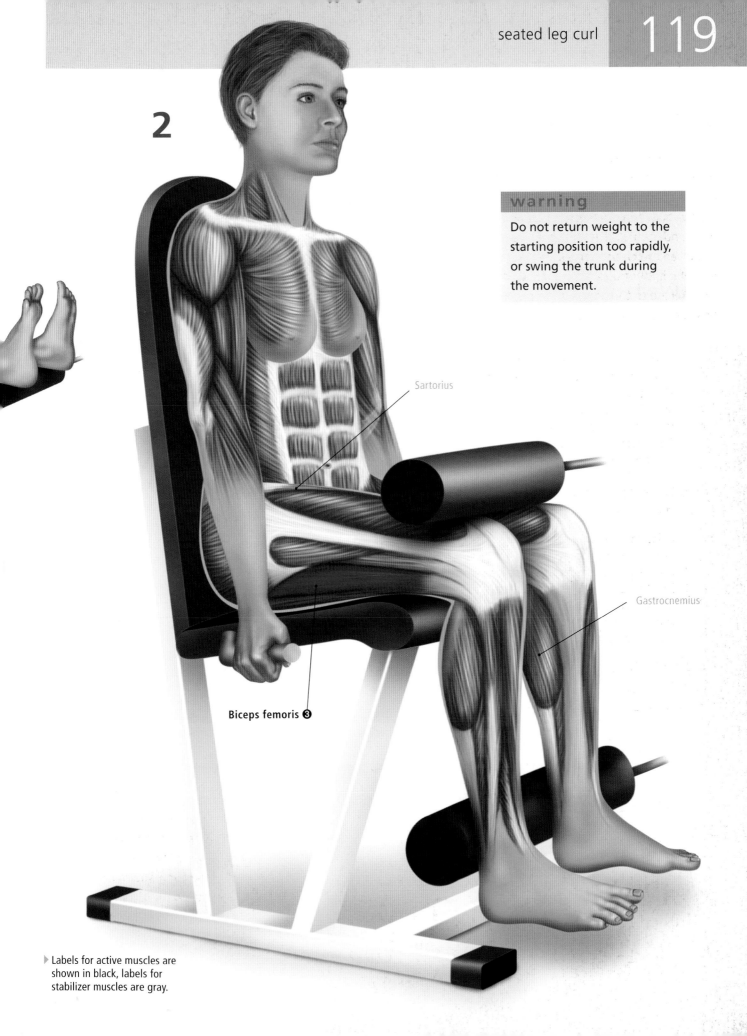

warning

Do not return weight to the starting position too rapidly, or swing the trunk during the movement.

Sartorius

Gastrocnemius

Biceps femoris ❸

▶ Labels for active muscles are shown in black, labels for stabilizer muscles are gray.

Lying Leg Curl

This popular exercise really works the semitendinosus, semimembranosus, and biceps femoris in the hamstring muscle group. Targeting this area assists running and kicking movements, which make the lying leg curl a good inclusion for most sports-specific programs. It is suitable for beginners and advanced exercisers. The standard version of this exercise requires a specialized machine, which restricts its execution to commercial gyms or well-equipped homes. However, the easy variation uses ankle weights, making it possible to perform the lying leg curl in a simple home environment. This variation is also a great inclusion for rehabilitation training programs that focus on the hamstrings.

1

2

how to

Lie facedown on the machine with knees slightly protruding from the edge of the padded surface, so knee joints line up with the rotation axis of the lever arm. Adjust pad behind legs to just above ankle level. Curl legs up as far as possible toward the buttocks. Take two or three seconds to slowly return to the starting position with legs extended. Keep upper body in contact with pads at all times.

variations

EASY

Use ankle weights and lie facedown on a sturdy bench or table, with the knees just protruding from the edge. Perform movement as per the standard exercise. This variation is not restricted to a gym setting, but it does limit how much weight can be lifted.

HARD

Target the eccentric phase of the movement, one leg at a time. Perform the exercise initially as per the standard movement, curling both legs toward the buttocks. Then, on the return phase, resist the pull of the weight with only one leg. Complete the desired number of eccentric overload repetitions on each leg.

active muscles

❶ Biceps femoris

❷ Semimembranosus and Semitendinosus

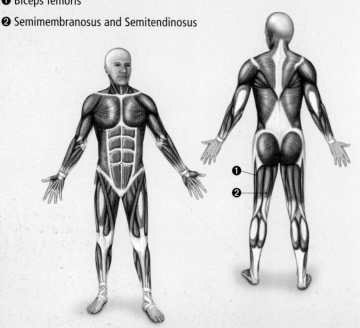

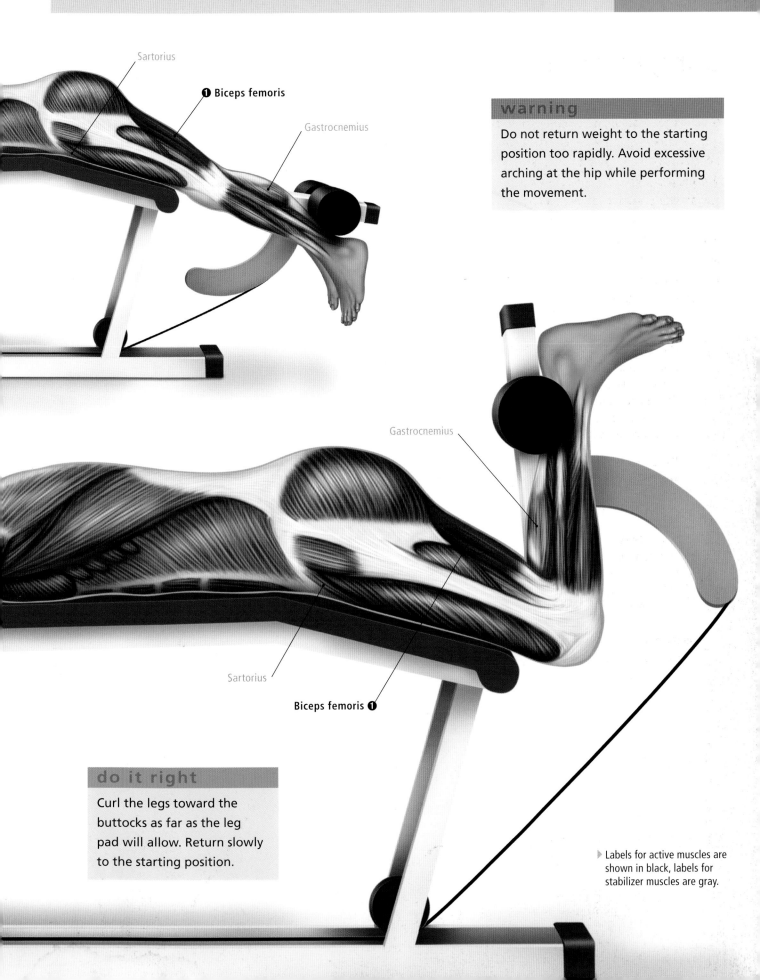

Sartorius

❶ Biceps femoris

Gastrocnemius

Gastrocnemius

Sartorius

Biceps femoris ❶

▶ Labels for active muscles are shown in black, labels for stabilizer muscles are gray.

Leg Press

This standard exercise is a common choice for targeting the quadriceps and gluteus maximus muscles. It is a popular alternative for trainers who prefer not to use the squat exercise. The leg press is beneficial to any sports that involve running, jumping, and kicking. It has both seated and lying variations, and is suitable for both beginners and advanced trainers. Although the leg press requires a relatively expensive machine, it does permit the use of heavy weights for more advanced lifters. For beginners, the focus is on using correct technique. The following explanation centers on the 45-degree leg press, which is a standard piece of equipment in most gyms.

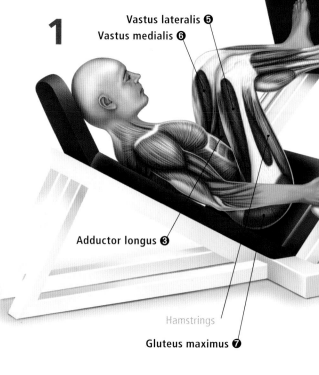

1

Vastus lateralis ❺
Vastus medialis ❻

Adductor longus ❸

Hamstrings

Gluteus maximus ❼

how to

Assume a seated, reclined position and place feet on the baseplate above, toes pointing forward. Apply pressure to the baseplate, to "take" the weight, then release the safety catches. Slowly lower the weight toward the body as far as is comfortable. Push legs back out until fully extended. After completing the desired number of repetitions, replace safety catches and slowly relax the pressure on the baseplate.

variations

EASY

Perform the standard exercise, but only lower the baseplate through the initial quarter of the range of motion. As strength builds, be sure to progress to the standard exercise because this variation does not train the muscles through the full range of motion.

HARD

Choose a weight that is light enough to slowly lower with one leg in a controlled movement, but heavy enough that it requires both legs to raise it again. The enhanced eccentric load strengthens muscles by increasing the stress on the muscle fibers to a greater level than is possible if both legs perform the lowering movement together. Perform sets on each leg.

active muscles

❶ Adductor brevis (under Adductor longus)

❷ Vastus intermedius (under Rectus femoris)

❸ Adductor longus

❹ Adductor magnus

❺ Vastus lateralis

❻ Vastus medialis

❼ Gluteus maximus

2

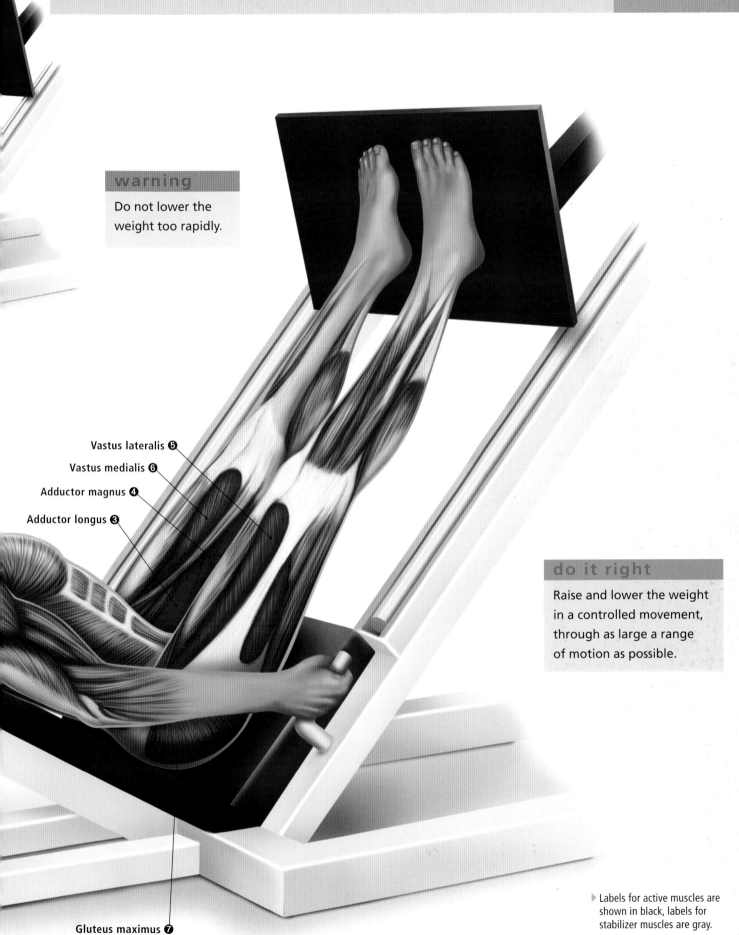

Do not lower the
weight too rapidly.

Vastus lateralis ❺
Vastus medialis ❻
Adductor magnus ❹
Adductor longus ❸

do it right

Raise and lower the weight
in a controlled movement,
through as large a range
of motion as possible.

Gluteus maximus ❼

▶ Labels for active muscles are
shown in black, labels for
stabilizer muscles are gray.

Nordic Hamstrings

This hamstring exercise is specifically aimed at developing eccentric strength of the hamstring muscle group. Use it for both general strengthening programs and in rehabilitation settings. The Nordic hamstring exercise is also a popular choice for sports that have a high incidence of hamstring injuries—the main culprits being football and track sprinting, because of the eccentric contribution just prior to and during foot strike in the running motion. It is suitable for beginners through to advanced exercisers—match the resisted range of motion and external weight to the training status of the individual.

how to

Kneel on a flat, padded surface, such as a sit-up mat, with back straight, hands at chest level, and palms facing away from the body. Have a spotter hold the lower legs down. Slowly lower body toward the ground in a controlled movement, resisting the urge to fall by contracting the hamstring muscles. Do not bend at the hips. At ground level, use a push-up motion to return to the starting position. Perform three to five repetitions per set.

variations

EASY

Resist with the hamstrings only until the controlled downward motion cannot be contained. At that point, relax hamstrings and fall freely to the ground, slowing the descent with arms and hands. Complete the desired number of repetitions.

HARD

Add extra weight to the body by holding a weight plate or dumbbell across the chest, then perform as per the standard exercise. The extra weight increases the stress on the hamstring muscles, allowing progressive overload to be incorporated into the training program.

active muscles

❶ Semitendinosus
❷ Biceps femoris
❸ Semimembranosus

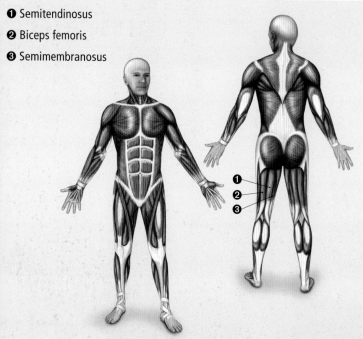

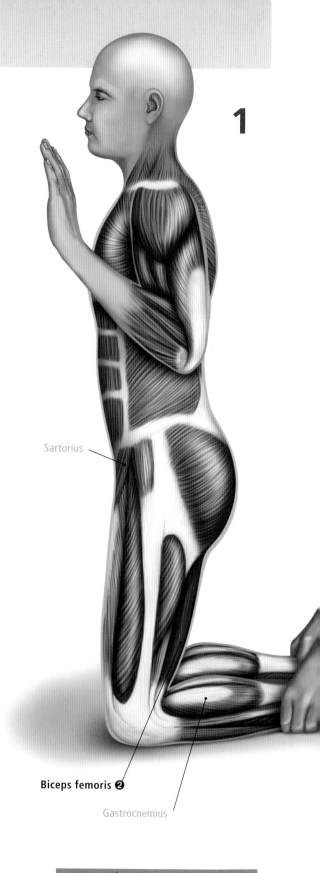

1

Sartorius

Biceps femoris ❷

Gastrocnemius

warning

If experiencing excessive stress on hamstring muscles, relax at that point and use arms and hands to break the fall.

2

do it right

Slowly lower as far as possible. Avoid jerky movements and do not bend at the hips.

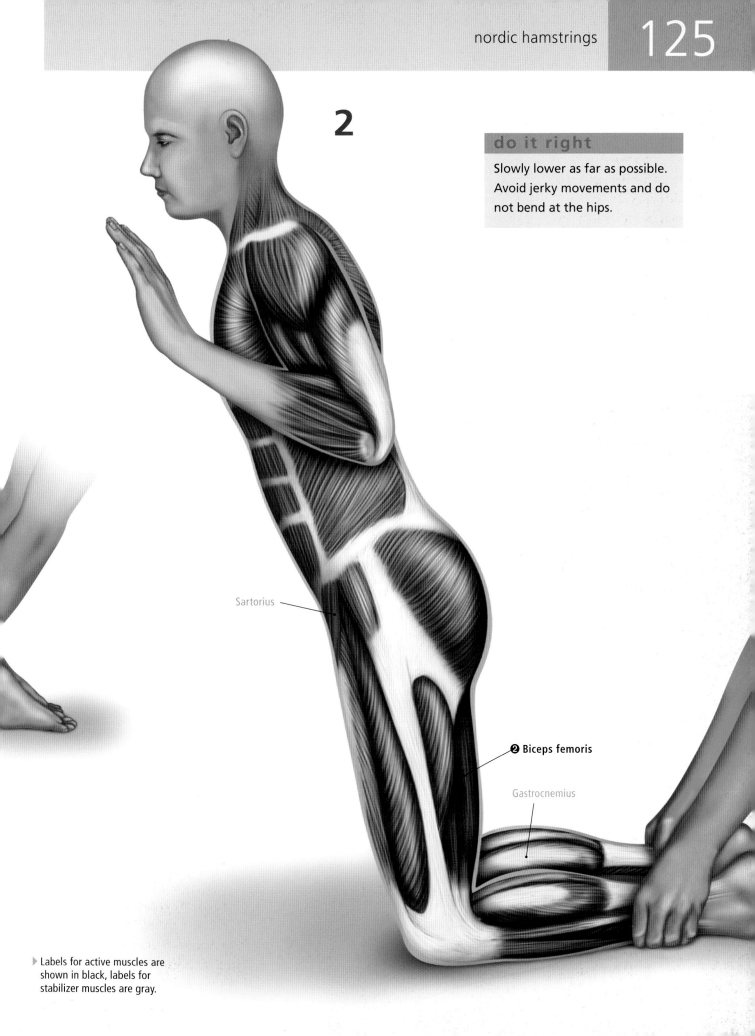

Sartorius

❷ **Biceps femoris**

Gastrocnemius

▶ Labels for active muscles are shown in black, labels for stabilizer muscles are gray.

Hip Adduction

This strengthening exercise is too often ignored, especially by male trainers. It provides a great workout for the adductor muscles, which pull the legs together (adduction) and allow flexion, extension, and medial or lateral rotation at the hip. This makes it a clever choice for football and many court sports, which involve cutting maneuvers and crossover steps. Depending on the variation chosen, it can be suitable for both home and gym environments and is appropriate for beginners through to advanced trainers. The adductors are not strong muscles, so choose a light weight when first performing this exercise and progressively overload as muscles become stronger.

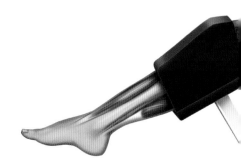

how to

Sit in a hip-adduction machine, with padded supports lying along the inside of the legs. Choose and set a starting position with legs open as wide as comfortably possible. Pull the padded supports together until they touch, by pushing against them with the legs. Slowly open legs, resisting against the pull of the weight return to the starting position.

variations

EASY

Lie on the ground and push one leg at a time against a stability ball that is fixed against an immovable object, such as the other leg or a wall. Keeping the leg straight, try to depress the ball as much as possible. Slowly resist the expansion of the ball as the leg moves outward to return to the starting position. Repeat on alternate leg.

HARD

Stand side-on to a standing pin-loaded weight stack, legs 3 feet (91 cm) apart. Attach a low pulley strap just above the ankle of the leg closest to the machine. The leg should be straight and raised to the side, 6 inches (15 cm) off the ground, with the other leg firmly on the floor. Slowly pull the raised leg toward the ground-based leg until they meet. Return to the starting position. Repeat on alternate leg.

active muscles

❶ Pectineus

❷ Adductor brevis
(under Adductor longus)

❸ Adductor longus

❹ Adductor magnus

❺ Gracilis

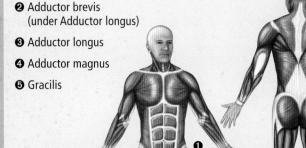

do it right

Control the speed of inward and outward movements. Keep legs straight while performing the exercise.

warning

Do not use excessive weight, as adductor muscles cannot tolerate the same loads as muscles such as the quadriceps.

▶ Labels for active muscles are shown in black, labels for stabilizer muscles are gray.

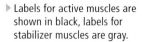

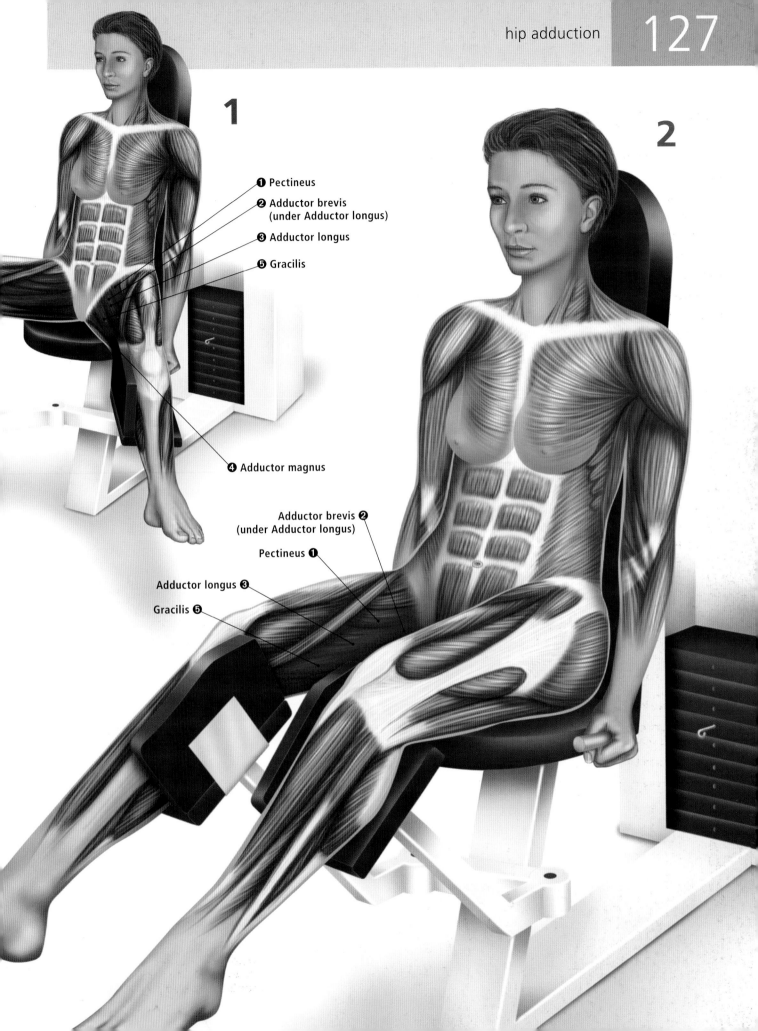

1

❶ Pectineus

❷ Adductor brevis
(under Adductor longus)

❸ Adductor longus

❺ Gracilis

❹ Adductor magnus

2

Adductor brevis ❷
(under Adductor longus)

Pectineus ❶

Adductor longus ❸

Gracilis ❺

Hip Abduction

This often-overlooked exercise, especially by male trainers, is effective in working the hip abductor muscles that move the legs apart. It is a good choice to balance exercise programs, as it improves weakness of the gluteus medius muscles. This also makes it a popular choice for sports-specific training and occupational rehabilitation programs. The abductors are not a strong muscle group, so take care choosing weight when starting to train the area. This exercise is suitable for home and gym environments, depending on the variation, and is appropriate for all levels of trainers.

how to / variations

EASY

Sit in the hip-abduction machine, with the padded supports lying along the outside of the legs. Choose and set a starting position with legs closed. Push the padded supports apart as far as possible, then resist against the weight as legs slowly close to return to the starting position. Complete the desired number of repetitions.

Place a stability ball against an immovable surface, such as a wall. Stand or lie with one leg against the ball and feet close together. Hold the supporting leg still, and apply pressure to the ball with the other leg. Squash the ball against the wall as legs open, keeping both legs straight. During the return phase of the motion, slowly resist the expansion of the ball as legs are pushed back together. Repeat on alternate leg.

HARD

Stand side-on to a standing pin-loaded weight stack. Attach a low pulley strap just above the ankle of the leg farthest from the machine. Both legs should be straight, feet on the ground. Slowly raise the leg with the attached strap outward, away from the body, until the foot is about 2 feet (61 cm) off the ground. Slowly return foot to the ground, resisting the pull of the weight. Repeat on alternate leg.

active muscles

❶ Tensor fasciae latae
❷ Piriformis
❸ Gluteus medius
❹ Gluteus minimus

(2, 3, and 4 all under Gluteus maximus)

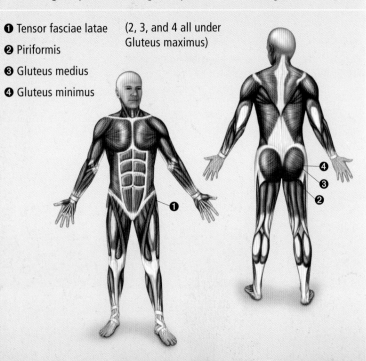

do it right

Keep legs straight throughout the exercise. Resist against the pull of the weight stack and return legs slowly to the closed position.

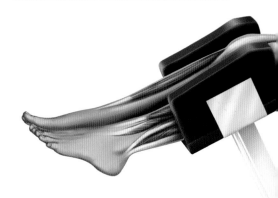

warning

Use a weight that allows a smooth, controlled motion. Maintain correct technique by avoiding any twisting of the hip.

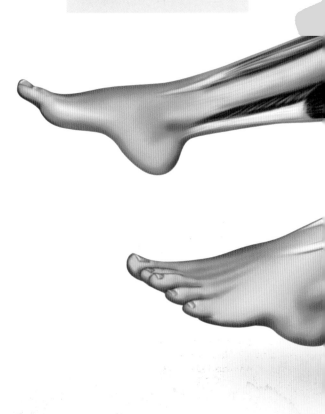

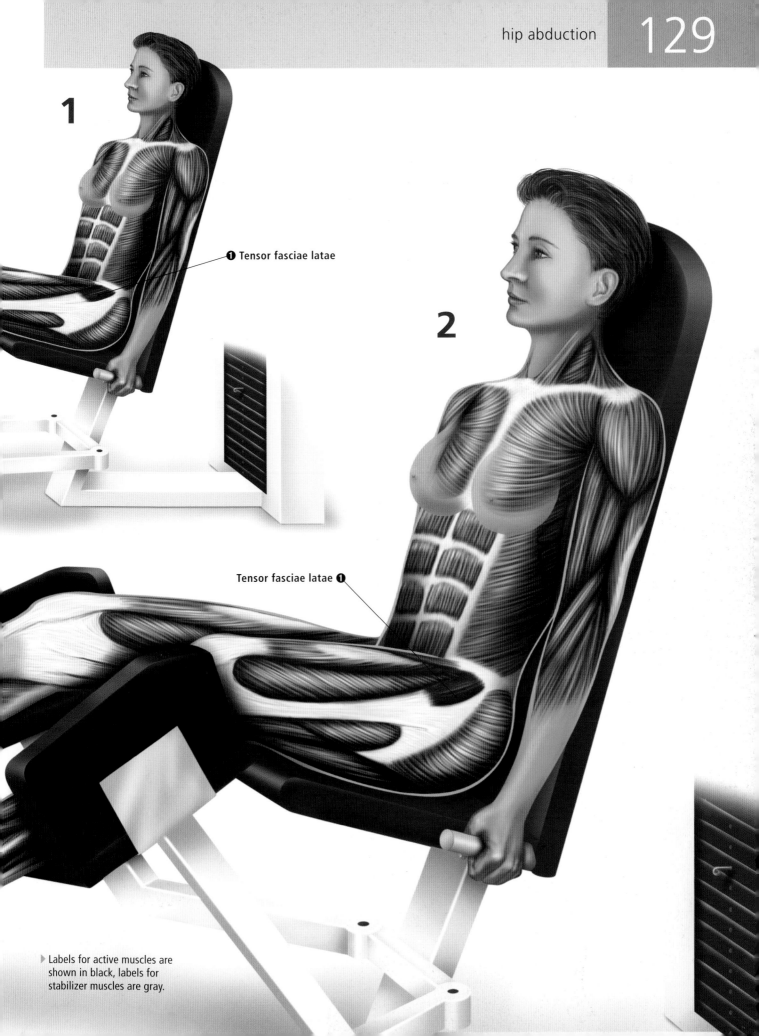

1

❶ Tensor fasciae latae

2

Tensor fasciae latae ❶

▶ Labels for active muscles are shown in black, labels for stabilizer muscles are gray.

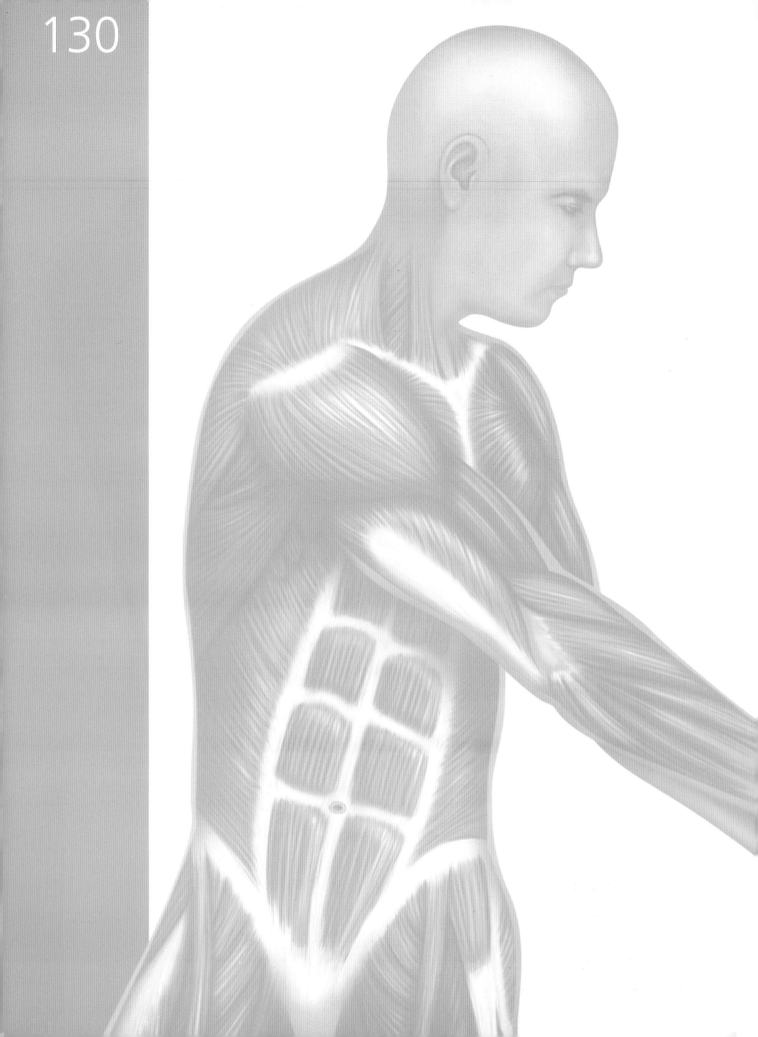

Trunk Exercises

The core, a vitally important area for strength, fitness, and stability, is the focus of exercises in this section. Do not mistake the term "core stability" for core strength. Few people truly require strong core muscles—it is endurance, tone, and the ability to support a stable posture while completing other activities that is most important.

All exercisers will benefit from core stability and lower back endurance. Together, they protect against lower back pain—one of the most common complaints in Western society. For athletes, core stability provides a stable foundation from which to achieve longer and stronger throws, better absorb body contact, and maintain balance.

Correct technique is essential to ensure that trunk exercises are safe and effective. Maintain a neutral spine position and minimize the use of accessory muscles.

Plank

This core-stabilizing exercise is also used for scapular-stabilization training. Rather than building strength, the plank develops endurance and stability. In particular, it improves endurance in the lower back, abdominal, shoulder, and hip muscles. The aim is to hold the position for as long as possible, depending on fitness levels. The current world-record holder maintained the plank position for more than one hour! This exercise is ideal for sports where athletes need to stay stable or rigid while performing a task, such as gymnastics and diving, or sports where body contact needs to be absorbed or countered. Use the plank in early to moderate shoulder-rehabilitation programs, and as a precursor to performing a push-up.

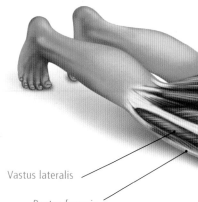

Vastus lateralis

Rectus femoris

External oblique ❸

Internal oblique ❷
(under External oblique)

Rectus abdominis ❶

how to

Start by lying facedown on the ground with elbows bent, hands curled into fists just in front of the shoulders, and feet on their toes. Push the whole body up so that it rests only on the forearms and toes. Keep eyes down, shoulders back, and ensure that the back is straight. The body should be in a flat plane from the head to the heels. Hold the position for a minimum of 15 seconds. Work to increase the amount of time the position is held.

variations

EASY

Keep knees on the ground rather than resting on the toes if lacking the upper body strength or endurance to hold the plank. Ensure that the body is held in a flat plane from the head to the knees.

HARD

Lift one foot off the ground for five seconds at a time. This works the oblique and gluteal muscles, as they engage to resist rotation and maintain the leg in the air. For an advanced plank workout, try the "side plank." Move from the left side, to center, to the right side, holding each position for 30–60 seconds.

2

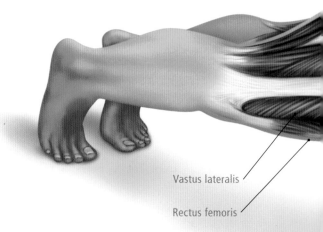

Vastus lateralis

Rectus femoris

active muscles

❶ Rectus abdominis

❷ Internal oblique
(under External oblique)

❸ External oblique

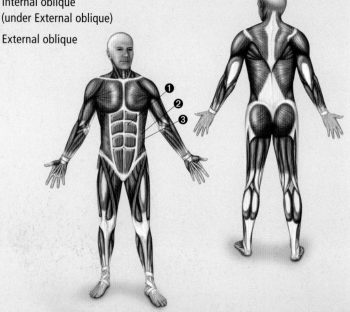

do it right

Use a mirror when first performing this exercise to ensure correct positioning.

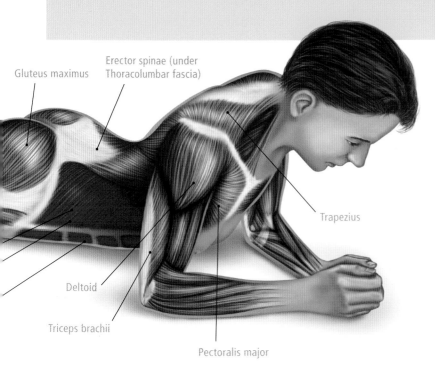

Gluteus maximus

Erector spinae (under Thoracolumbar fascia)

Trapezius

Deltoid

Triceps brachii

Pectoralis major

1

warning

Do not arch the lower back.
If abdominal muscles are weak,
start with the easy variation
and build up endurance.

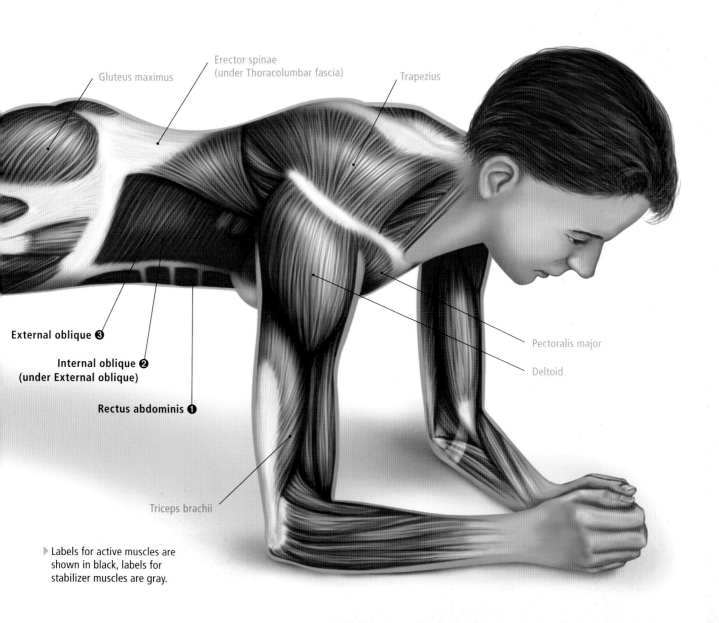

Gluteus maximus

Erector spinae
(under Thoracolumbar fascia)

Trapezius

Pectoralis major

Deltoid

External oblique ❸

**Internal oblique ❷
(under External oblique)**

Rectus abdominis ❶

Triceps brachii

▶ Labels for active muscles are
shown in black, labels for
stabilizer muscles are gray.

Crunch

This classic abdominal exercise has been performed thousands of times. However, a common misconception is that it assists in losing excess fat around the girth. This is not true, as the crunch is not a fat-burning exercise; rather, it is a strength and endurance exercise for the rectus abdominis, internal and external obliques, psoas, and iliacus. In the curled position, the lower back muscles are stretched and the gluteals, quadriceps, adductors, and hamstrings help stabilize the lower half of the body. The crunch does not require any special equipment, so it can be performed just about anywhere and is great for home exercise routines or trainers on the go.

how to

Lie back, with knees and hips flexed and feet flat on the floor. Cross arms over the chest with hands flat on opposite shoulders, or for a slightly harder option, place hands lightly beside the head. Squeeze and contract abdominal muscles to lift the head, shoulders, and scapula off the floor. Slowly lower back to the starting position.

variations

EASY

Use a spotter to hold the ankles or hook the feet under a solid stationary object, such as a bench or cupboard, if abdominals do not have the strength and power needed to perform this exercise without assistance. Contract through the abdominals and do not pull up with the feet or the legs, as this reduces the effectiveness of the exercise.

HARD

Reach arms back overhead or hold a weight overhead to really engage the abdominal muscles. These variations also make the lower limb stabilizers work harder to keep the feet down on the floor. However, ensure that the correct technique is not compromised.

active muscles

❶ Rectus abdominis

❷ Internal oblique
(under External oblique)

❸ External oblique

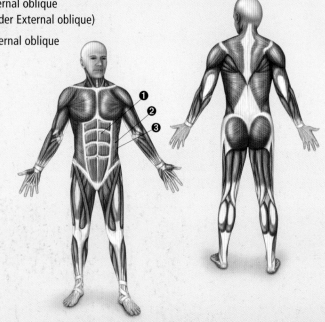

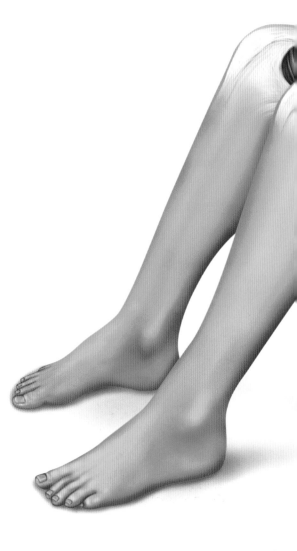

Rectus femoris

❶ Rectus abdominis

1

❸ External oblique

**❷ Internal oblique
(under External oblique)**

Gluteals

2

Rectus femoris

Rectus abdominis ❶

❸ External oblique

**❷ Internal oblique
(under External oblique)**

Gluteals

▶ Labels for active muscles are
shown in black, labels for
stabilizer muscles are gray.

Cross-body Crunch

This version of the exercise increases the challenge by requiring greater movement amplitude and adding a twisting component to the standard crunch. In turn, it targets a greater portion of internal and external oblique muscles and calls more on the secondary stabilizers. The cross-body crunch can be used as a progression toward the bicycle exercise, because the movement is similar but demands less from leg muscles and spinal stabilizers. Much like the standard crunch, this exercise is versatile and can be performed almost anywhere, as no special equipment is needed.

how to

Lie back with knees and hips flexed, feet flat on the floor, and hands placed lightly beside the head. Squeeze and contract the abdominals to lift the head, shoulders, and scapula off the floor while twisting toward the right side. At the same time, lift the left leg so that the right elbow and left knee touch over the chest. Slowly return to the starting position and repeat with alternate arm and leg.

variations

EASY

Drop the leg movement and just concentrate on lifting and twisting the upper half of the body if lacking the strength or endurance to execute the standard exercise in a smooth, controlled manner. This variation minimizes the abdominal work and decreases the importance of secondary stabilizers.

HARD

Perform the exercise on an incline bench, with head lower than the feet. This position increases the resistance by simply adding an extra gravitational pull. To increase the resistance even further, hold a weight behind the head.

do it right

Be sure to feel the "squeeze" when contracting the abdominals. Do not pull up with the arms or legs.

active muscles

❶ Rectus abdominis

❷ Internal oblique
   (under External oblique)

❸ External oblique

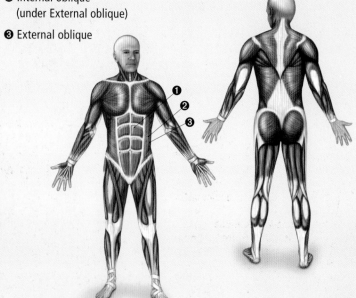

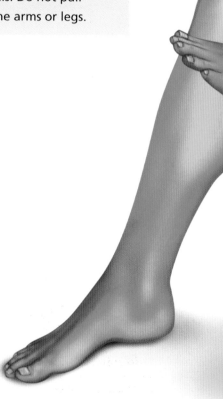

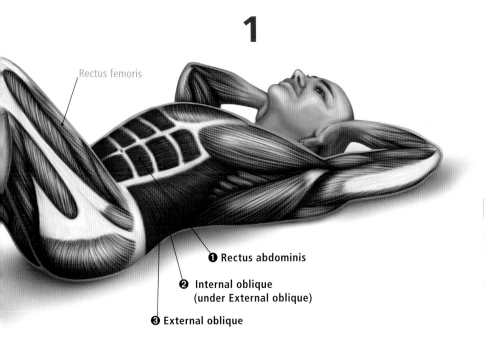

1

Rectus femoris

❶ Rectus abdominis

❷ Internal oblique
(under External oblique)

❸ External oblique

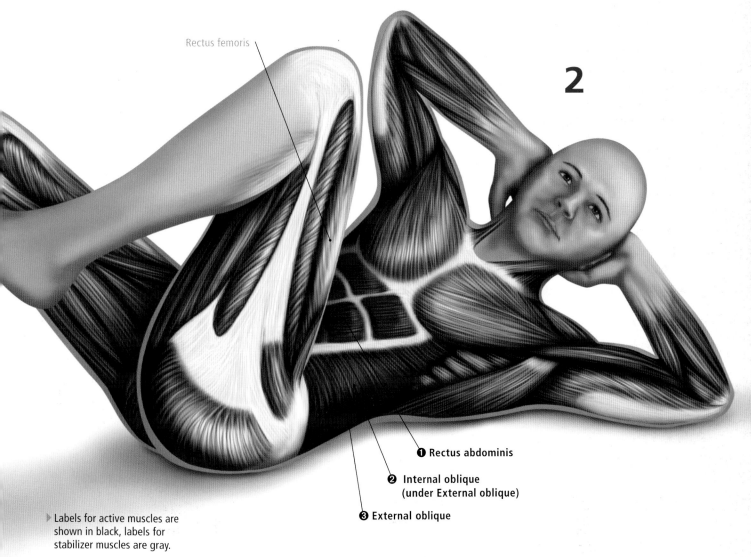

Rectus femoris

2

❶ Rectus abdominis

❷ Internal oblique
(under External oblique)

❸ External oblique

▶ Labels for active muscles are shown in black, labels for stabilizer muscles are gray.

Bicycle

This advanced exercise works the rectus abdominis hard. In fact, in a San Diego State University study that measured muscle activation, it ranked top of all common abdominal exercises for rectus activation. The bicycle also ranked highly for activation of the obliques, making it one of the most effective ways to target the superficial abdominal muscles. Definition in these muscles results in the "six-pack" look that so many trainers desire. Use the bicycle as an advanced core-stability exercise, but only once deep core muscles have been effectively trained. Maintaining a neutral spine while lifting and moving the legs places high demands on the muscles.

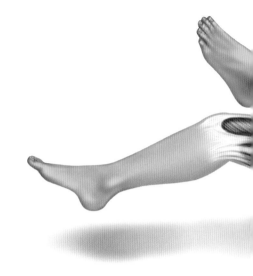

how to

Lie flat on the floor, with the lower back pressed to the ground, then contract the core muscles. Holding the head gently with the hands, lift the knees to a 45-degree angle. Move the legs through a bicycle pedal motion, and alternately touch the elbow to the opposite knee while twisting back and forth. Breathe evenly throughout the exercise.

variations

EASY

The most challenging part of the bicycle is maintaining stability when the leg is fully extended. Keep knees slightly flexed throughout the exercise if finding it hard to maintain correct form or a neutral spine position.

HARD

Try a "seated bicycle." Sit down and lean back slightly to make the core stabilizers start working even before any movement begins. Extend legs and lift both heels off the floor. Perform the bicycle pedal action and move opposite elbows to knees. To make this exercise even more challenging, touch the outside of the elbow to the outside of the knee.

do it right

Throughout the exercise, pull the deep abdominal muscles toward the spine and maintain a neutral spine position.

active muscles

❶ Rectus abdominis
❷ Internal oblique
   (under External oblique)
❸ External oblique

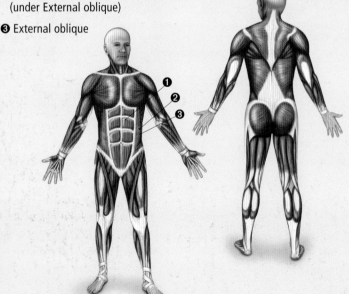

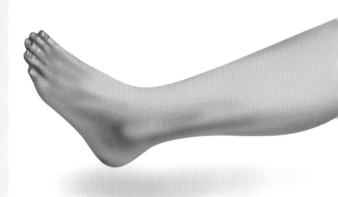

▶ Labels for active muscles are shown in black, labels for stabilizer muscles are gray.

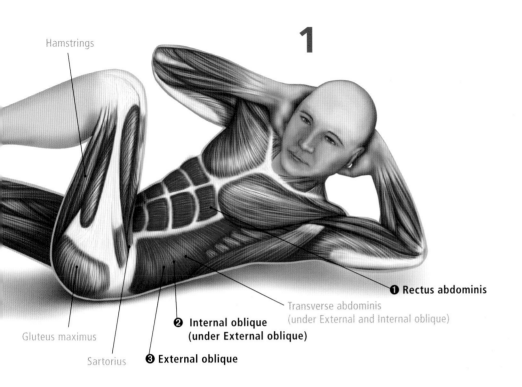

1

Hamstrings

❶ Rectus abdominis

Transverse abdominis
(under External and Internal oblique)

❷ Internal oblique
(under External oblique)

❸ External oblique

Gluteus maximus

Sartorius

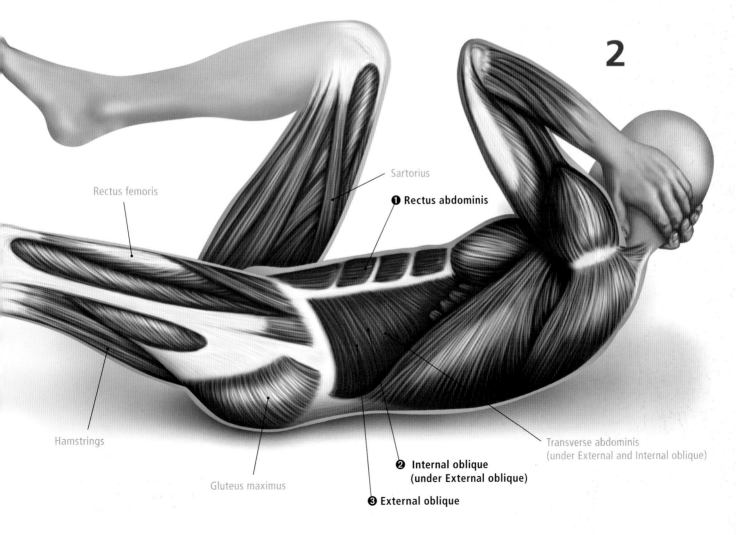

2

Rectus femoris

Sartorius

❶ Rectus abdominis

Hamstrings

Gluteus maximus

❷ Internal oblique
(under External oblique)

❸ External oblique

Transverse abdominis
(under External and Internal oblique)

Superman

This core-stability exercise is also known as the "bird dog," as the position moves between resting on all fours and extending arms and legs. It works predominantly on the posterior core, which is made up of the obliques, erector spinae, multifidus, and gluteals. The superman develops strength and stability around the neck and shoulder girdle by improving extensor endurance and encouraging rotational stability. Good lumbar extensor endurance is essential to protecting against lower back pain, and rotational stability is important in many sports, especially bat and racket sports. The superman does not require any special equipment—it is a floor or mat exercise and, as such, can easily be part of a home or gym routine.

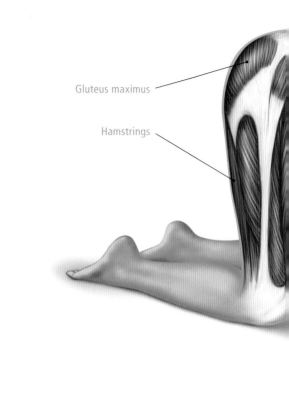

Gluteus maximus

Hamstrings

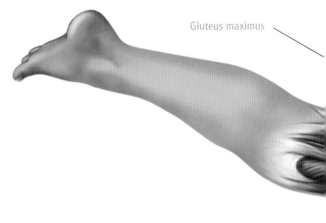

Gluteus maximus

Hamstrings

how to

Kneel on the ground with hands placed in front of the body, about shoulder-width apart. Tilt the pelvis back and forth to find the neutral spine position. In this position, brace the abdominals and lift one hand and the opposite knee just off the ground, balancing on the alternate hand and knee. Once comfortably stable, fully extend the arm and leg. Try to keep a flat plane all the way from the hand to the foot. Hold this position for about ten seconds. Return to the starting position and extend on the opposite side.

variations

EASY

Keep both hands on the ground and extend only one leg at a time if finding it hard to stabilize with both leg and arm extended.

HARD

Start on hands and toes, rather than on hands and knees, for an advanced workout. Shoulder stabilizers, core stabilizers, and hamstrings will all have to work harder because of the longer segment being stabilized and the decreased surface contact area in this variation.

active muscles

❶ Rectus abdominis

❷ Internal oblique (under External oblique)

❸ External oblique

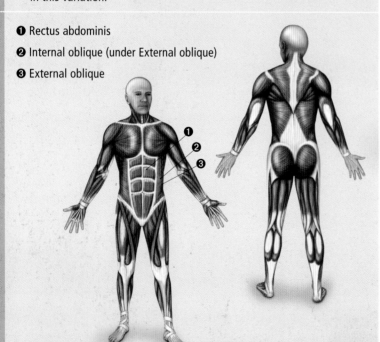

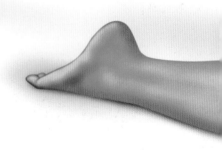

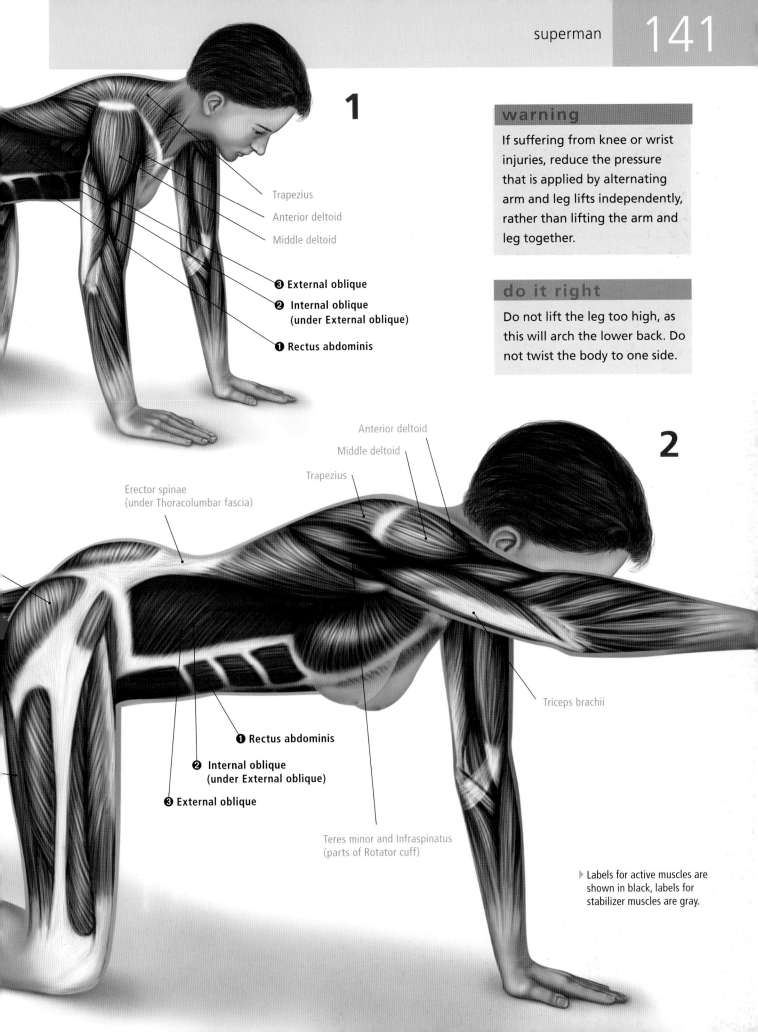

1

Trapezius

Anterior deltoid

Middle deltoid

❸ External oblique

❷ Internal oblique
(under External oblique)

❶ Rectus abdominis

warning

If suffering from knee or wrist injuries, reduce the pressure that is applied by alternating arm and leg lifts independently, rather than lifting the arm and leg together.

do it right

Do not lift the leg too high, as this will arch the lower back. Do not twist the body to one side.

2

Anterior deltoid

Middle deltoid

Trapezius

Erector spinae
(under Thoracolumbar fascia)

Triceps brachii

❶ Rectus abdominis

❷ Internal oblique
(under External oblique)

❸ External oblique

Teres minor and Infraspinatus
(parts of Rotator cuff)

▷ Labels for active muscles are shown in black, labels for stabilizer muscles are gray.

Bridge

This strengthening exercise targets the gluteal, hamstring, abdominal, and lower back muscles. It is an excellent rehabilitation exercise for lower back pain and core stability. Poor endurance of the lower back muscles is one of the most dangerous risk factors in developing lower back pain. In response, the bridge has proven to be one of the most effective ways to activate and strengthen lower back muscles, such as the multifidus and the erector spinae. Use this exercise as a preventative measure, or as part of a training program when recovering from injury. There are many variations of the bridge, so exercisers can progress from easy to advanced forms. It requires little equipment and is therefore easy to perform in most settings.

1

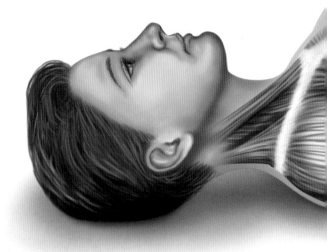

how to

Lie back with hands by sides, knees bent, and feet flat on the floor. Tighten the abdominal and gluteal muscles, push down through the heels, and raise the hips to create a straight plane from the knees to the shoulders. Keep the core muscles tight and do not let hips drop or the back arch. Hold this position for 20–30 seconds. Once the muscles begin to fatigue, slowly lower back to the starting position.

variations

EASY

Depending on individual fitness levels, lift the hips only a small way off the ground until body strength develops, or hold the standard position for a reduced amount of time.

HARD

Try a "one-legged bridge." Perform the exercise as outlined in the standard version but start with only one knee bent, holding the other leg fully extended. Push up with the bent leg and aim for a flat plane from the shoulders to the foot of the straight leg. Resist the urge to rotate the body; keep hips flat and do not twist to one side.

warning

Dropping suddenly from the raised position can jar the lower back and cause pain or injury.

active muscles

❶ Erector spinae
❷ Multifidus
  (under Erector spinae)
❸ Rectus abdominis
❹ Internal oblique
  (under External oblique)
❺ External oblique
❻ Gluteus maximus

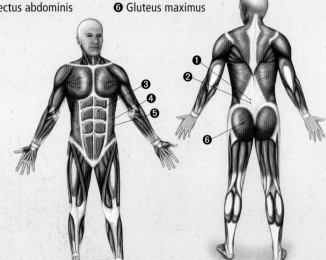

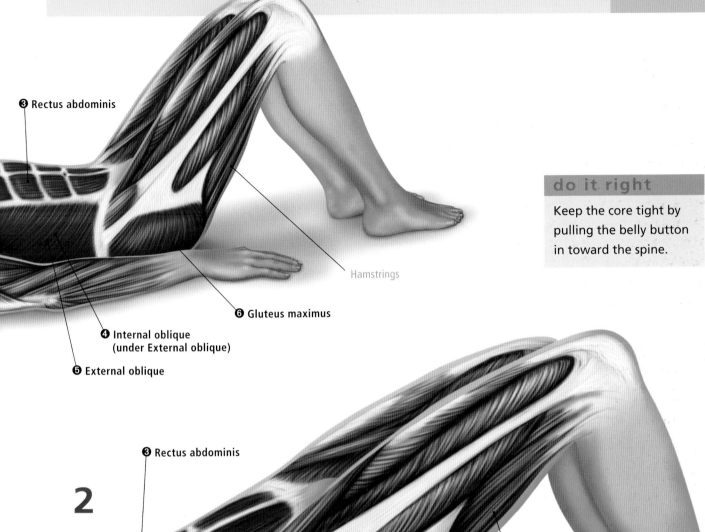

❸ Rectus abdominis

Hamstrings

do it right

Keep the core tight by pulling the belly button in toward the spine.

❻ Gluteus maximus

❹ Internal oblique
(under External oblique)

❺ External oblique

2

❸ Rectus abdominis

Hamstrings

❻ Gluteus maximus

❹ Internal oblique
(under External oblique)

❺ External oblique

▷ Labels for active muscles are shown in black, labels for stabilizer muscles are gray.

Back Extension

This bodyweight exercise both stretches and strengthens the muscles of the lower back. It is usually performed in the gym on a back-extension machine. In the lowered position, erector spinae, quadratus lumborum, and the thoracolumbar fascia are stretched. Then, as the body is raised, the back muscles and hamstrings must contract. Use the back extension to complement crunches or sit-ups and balance the mid-section. Many people perform abdominal exercises and upper back exercises but neglect the muscles in the lower back. Remember, strength and endurance in these muscles are essential to protect against developing lower back pain.

warning

Even the easy variation of this exercise puts strain on the lower back. Discontinue if experiencing any pain while performing the movement.

how to

Use a back-extension machine and position the pads to sit underneath the thighs and on top of the calves. Adjust the length to support the hips and pelvis, but ensure that none of the upper body is supported. Start in a neutral or slightly hyperextended position, then lower the body, bending at the waist. Lower as far down as back and hamstring flexibility will allow. Lift up and return to the starting position.

variations

EASY

Lie facedown on the floor instead of using a back-extension machine. Place hands behind the back and arch up to lift the head, neck, shoulders, and chest off the ground. This variation still works through the same range of muscles as the standard form, without relying on a back-extension machine.

HARD

Once comfortably able to complete three sets of 10–12 repetitions lifting your own bodyweight, try a weighted back extension. Take a plate from the barbell racks and hold it across your chest as you perform the movement. For safety, place the plate down on the ground at the end of the set, before dismounting the machine.

active muscles

❶ Erector spinae

❷ Multifidus

❸ Quadratus lumborum

(1, 2, and 3 all under Thoracolumbar fascia)

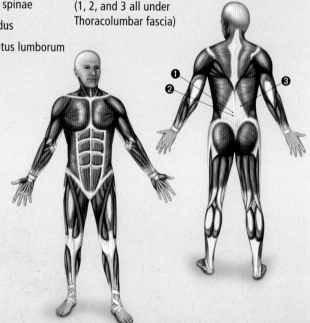

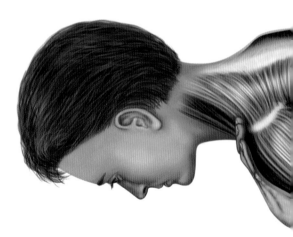

2

do it right

Do not swing or bounce at end positions; use a continuous, smooth motion.

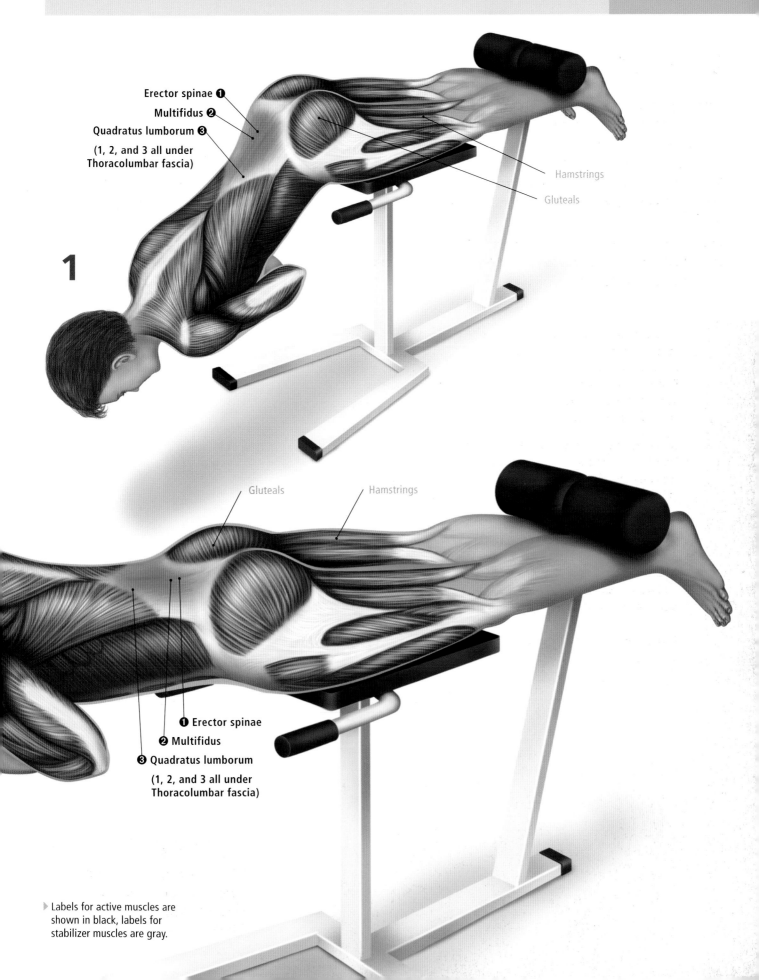

1

Erector spinae ❶
Multifidus ❷
Quadratus lumborum ❸
(1, 2, and 3 all under
Thoracolumbar fascia)

Hamstrings

Gluteals

Gluteals

Hamstrings

❶ Erector spinae
❷ Multifidus
❸ Quadratus lumborum
(1, 2, and 3 all under
Thoracolumbar fascia)

▶ Labels for active muscles are
shown in black, labels for
stabilizer muscles are gray.

Lift

1

This proprioceptive neuromuscular facilitation (PNF) exercise incorporates complex multidirectional movements that involve both the upper and lower limbs. It is often used in conjunction with the chop. The PNF lift comprises extension and rotation movements that strengthen torso rotation, the upper back, chest, shoulders, and arms. It can be performed with cable-weight machines, elastic-resistance bands, or a medicine ball. As with all core stability exercises, the aim is not to lift the heaviest weight, but to maintain perfect form throughout the set. Add the lift to later stages of injury rehabilitation programs and use it for sports-specific training. This exercise is perfect for racket and bat sports, especially golf.

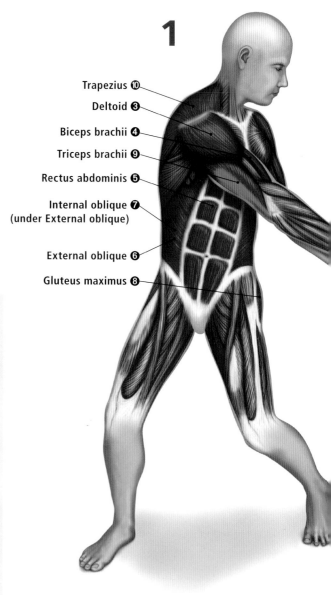

Trapezius ❿
Deltoid ❸
Biceps brachii ❹
Triceps brachii ❾
Rectus abdominis ❺
Internal oblique ❼ (under External oblique)
External oblique ❻
Gluteus maximus ❽

how to

The starting position can vary from kneeling, half kneeling, sitting on a stability ball, or standing. However, the movement pattern is always the same: from low to high; lifting with both hands; beginning with a pulling motion that moves across the mid-line of the body and ending with a pushing motion. For a standing lift, stand side-on or slightly forward and grasp the cable handle with both hands on one side of the body, below knee height. Lift with the upper or crossed-over arm dominating, twist and extend with the trunk, then push through and up with the other arm so that the movement ends on the other side of the body, hands above shoulder height. Slowly lower to the starting position. Perform sets from right to left, then left to right.

variations

EASY Stand with a wide or split stance to provide a greater base of support for the body and achieve more stability from the legs.

HARD Stand on one leg to decrease the stability supplied by the legs and put the entire focus of the exercise on the core muscles.

active muscles

❶ Erector spinae
❷ Multifidus (under Erector spinae)
❸ Deltoid
❹ Biceps brachii
❺ Rectus abdominis
❻ External oblique
❼ Internal oblique (under External oblique)
❽ Gluteus maximus
❾ Triceps brachii
❿ Trapezius

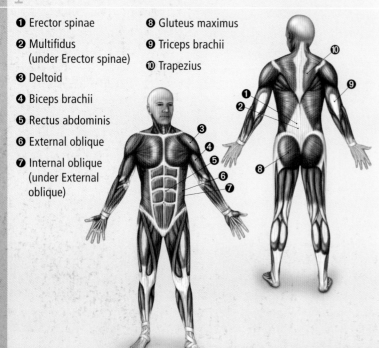

warning

Do not try to progress this exercise too quickly. It is all about form and technique. Too much weight too quickly can lead to injury.

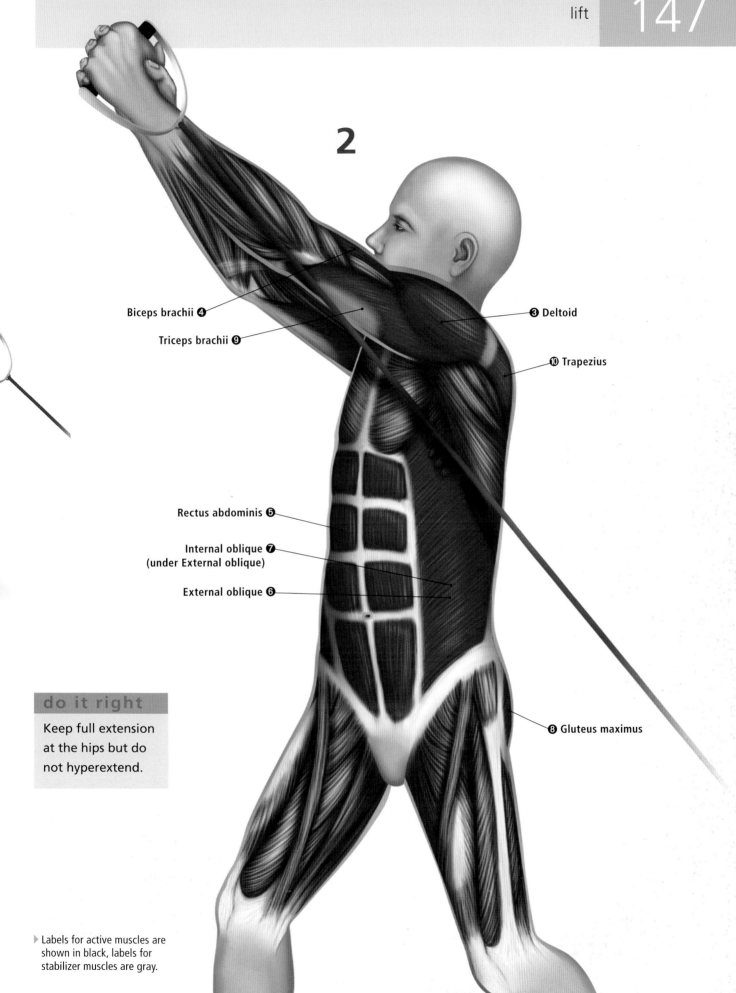

2

Biceps brachii ❹

Triceps brachii ❾

❸ Deltoid

❿ Trapezius

Rectus abdominis ❺

Internal oblique ❼
(under External oblique)

External oblique ❻

❽ Gluteus maximus

do it right
Keep full extension
at the hips but do
not hyperextend.

▶ Labels for active muscles are
shown in black, labels for
stabilizer muscles are gray.

Chop

This companion exercise to the lift is essentially a mirror image of that exercise, this time moving from a high to a low position. Because of the mechanical advantage that the chop has, most exercisers will be able to lift about one-third more weight in this exercise compared to the lift. When combined with left and right, the chop-lift combination targets four core quadrants. A skilled physical therapist can specifically test for strength in each quadrant, to determine whether a client should target one area first or train all four immediately.

how to

The starting position can vary between kneeling, sitting, and standing. Grasp the handle with one palm facing down and one palm facing up. The chop movement is always from high to low, beginning with a pulling motion, where the crossed-over arm moves across the mid-line of the body, and ending with a pushing motion from the following arm. Choose a starting weight that will cause fatigue or loss of the stable position after 6–12 repetitions. Try to correct the position and to keep going when starting to lose the stable base. If unable to maintain form and posture, then finish the set. Repeat from the opposite side.

variations

EASY

Stand with feet wide apart and knees slightly bent. The greater the stability supplied by the legs, the less that is required from the core. Use elastic tubing to perform this exercise at home or on the road, to continue resistance training when unable to get to the gym.

HARD

Kneel on a stability ball for a real challenge! Using the ball incorporates an unstable surface, and by adding rotational movements, only those with extremely high levels of core stability will be able to keep this variation under control.

active muscles

❶ Deltoid
❷ Biceps brachii
❸ Rectus abdominis
❹ Internal oblique (under External oblique)
❺ External oblique
❻ Gluteus maximus
❼ Multifidus (under Erector spinae)
❽ Erector spinae
❾ Triceps brachii
❿ Trapezius

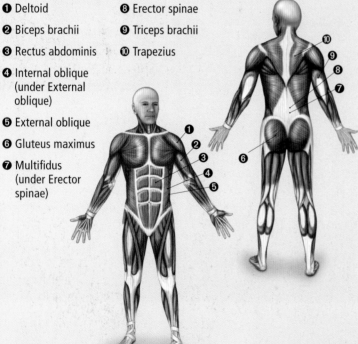

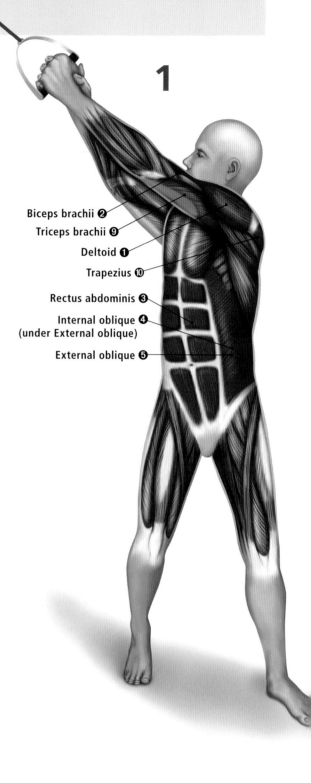

Biceps brachii ❷
Triceps brachii ❾
Deltoid ❶
Trapezius ❿
Rectus abdominis ❸
Internal oblique ❹ (under External oblique)
External oblique ❺

1

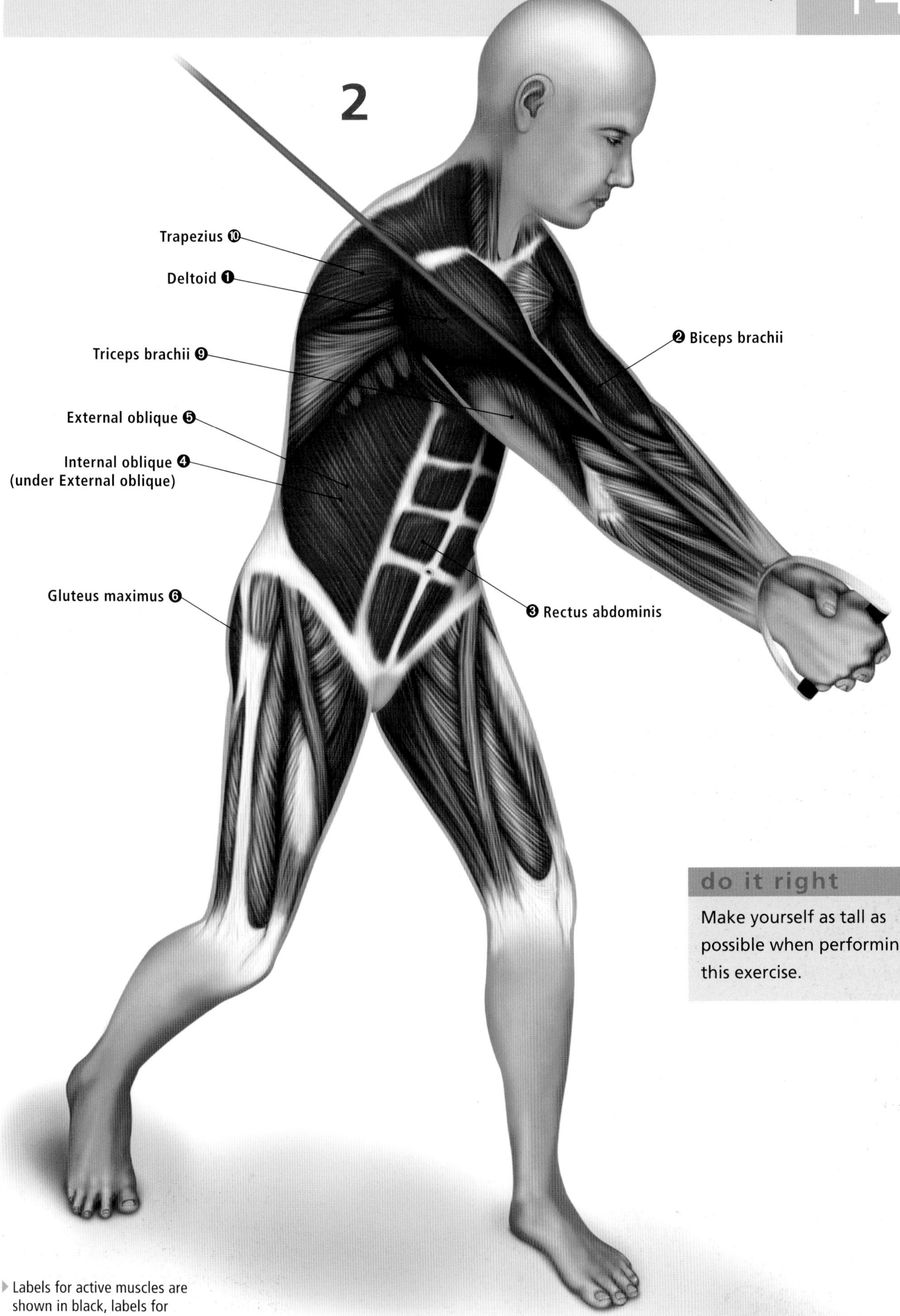

2

Trapezius ⑩

Deltoid ❶

❷ Biceps brachii

Triceps brachii ⑨

External oblique ❺

Internal oblique ❹
(under External oblique)

Gluteus maximus ❻

❸ Rectus abdominis

do it right

Make yourself as tall as
possible when performing
this exercise.

▶ Labels for active muscles are
shown in black, labels for
stabilizer muscles are gray.

Walkout

This stability-ball exercise works both the abdominal and back muscles. Using a stability ball forces multiple muscle groups to activate as the body struggles to remain balanced on the ball, and is a simple way to increase the challenge of many exercises. During the walkout, the abdominal and lower back muscles are active isometrically, while the shoulder and scapular stabilizers must maintain stability at the glenohumeral joint. Include this exercise to train core stability or rigidity, as well as to strengthen and stabilize the shoulder capsule. It is suitable as a general toning exercise, part of sport-specific training, or as an element of a rehabilitation program for back, neck, or shoulder injuries.

warning

Move to more challenging forms only when able to keep the ball completely stable in the full walkout position.

do it right

Do not let the lower back or hips sag when walking out. Keep the body rigid like a plank.

Hamstrings

Quadriceps femoris

2

how to

Lie on the stomach over a stability ball so that hands and toes can reach the floor. Place hands on the ball underneath the shoulders and contract the abdominal muscles. Lift feet off the floor and extend the legs so that the torso and legs are in a flat, horizontal plane. Keep the legs rigid and slowly walk hands forward. The ball will roll slightly, but keep walking out until only the feet rest on top of the ball. Slowly walk back and return to the starting position, with the stomach resting on top of the ball.

variations

EASY
The challenge of this exercise increases the further an exerciser walks out over the ball. Stop when thighs or knees are on top the ball if unable to keep stable walking all the way out to the feet.

HARD
For an extra stability challenge, lift one foot off the ball at the end of the walkout. Keep legs fully extended and do not let the ball roll from side-to-side underneath the foot. For an extra strength challenge, add a push-up when in the full walkout position. To increase the challenge even further, combine both variations.

active muscles

❶ Erector spinae
(under Thoracolumbar fascia)

❷ Rectus abdominis

❸ Internal oblique
(under External oblique)

❹ External oblique

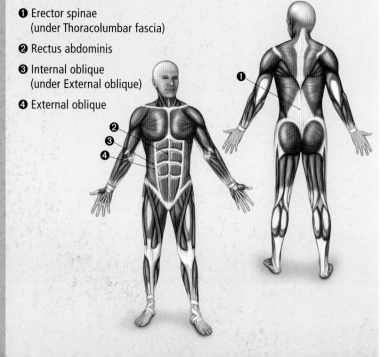

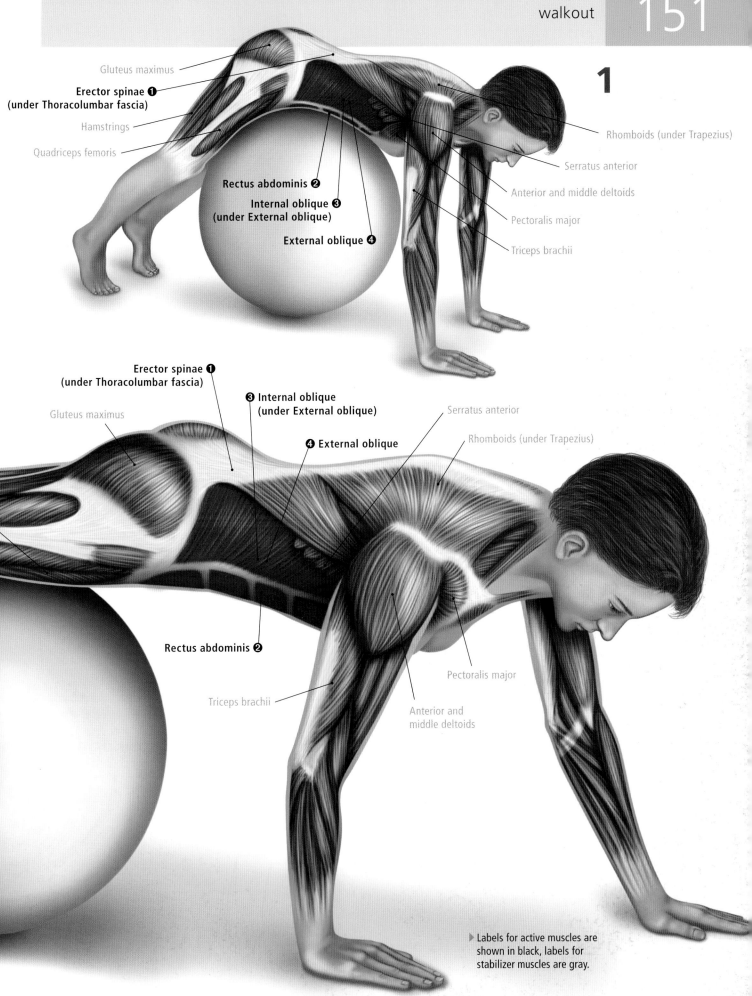

1

Gluteus maximus

Erector spinae ❶
(under Thoracolumbar fascia)

Hamstrings

Quadriceps femoris

Rectus abdominis ❷

Internal oblique ❸
(under External oblique)

External oblique ❹

Rhomboids (under Trapezius)

Serratus anterior

Anterior and middle deltoids

Pectoralis major

Triceps brachii

Erector spinae ❶
(under Thoracolumbar fascia)

Gluteus maximus

❸ Internal oblique
(under External oblique)

❹ External oblique

Serratus anterior

Rhomboids (under Trapezius)

Rectus abdominis ❷

Triceps brachii

Pectoralis major

Anterior and
middle deltoids

▶ Labels for active muscles are
shown in black, labels for
stabilizer muscles are gray.

Coloring Workbook

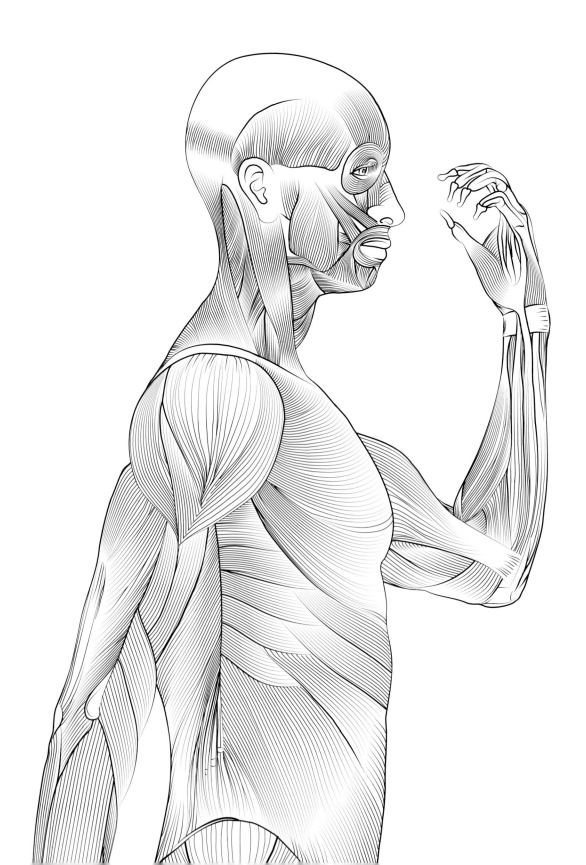

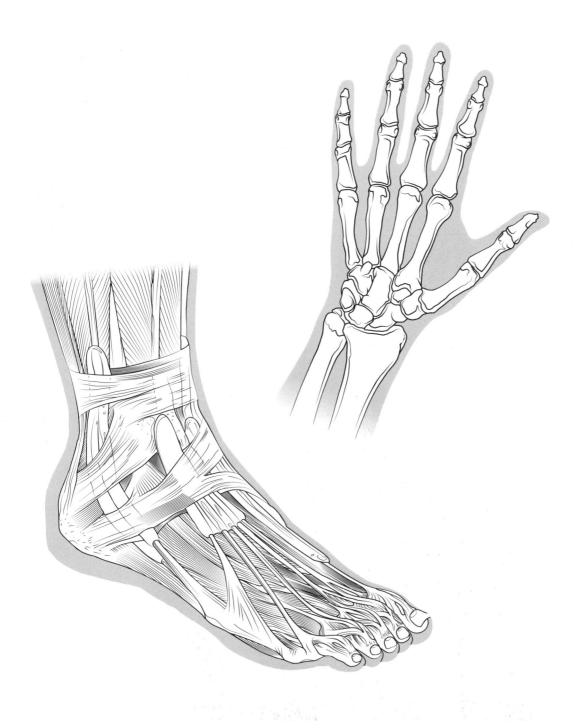

Muscular System

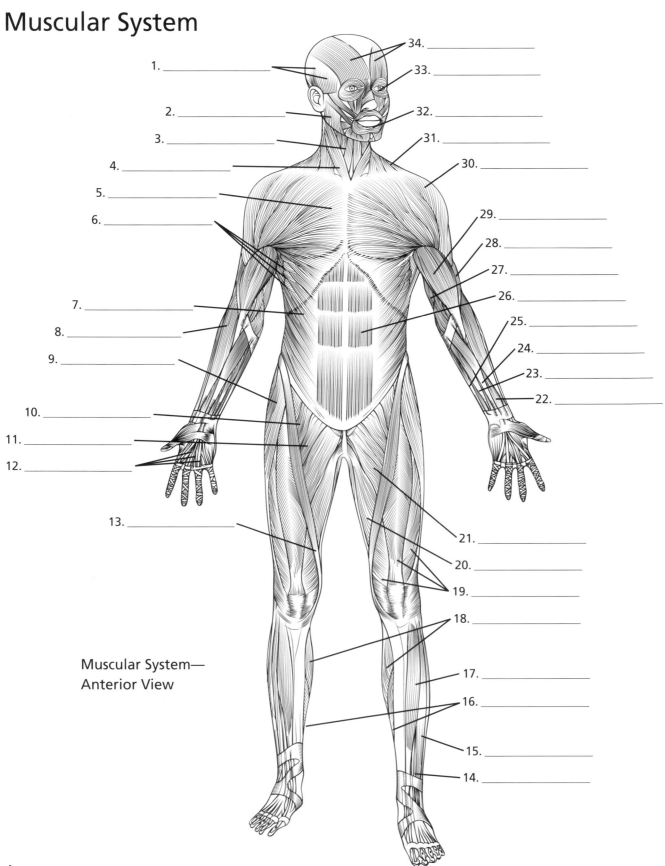

1. _____
2. _____
3. _____
4. _____
5. _____
6. _____
7. _____
8. _____
9. _____
10. _____
11. _____
12. _____
13. _____

14. _____
15. _____
16. _____
17. _____
18. _____
19. _____
20. _____
21. _____
22. _____
23. _____
24. _____
25. _____
26. _____
27. _____
28. _____
29. _____
30. _____
31. _____
32. _____
33. _____
34. _____

Muscular System—
Anterior View

Answers

1. Temporalis, 2. Masseter, 3. Sternohyoid, 4. Sternocleidomastoid, 5. Pectoralis major, 6. Serratus anterior, 7. External abdominal oblique, 8. Tensor fasciae latae, 9. Brachioradialis, 10. Iliopsoas, 11. Pectineus, 12. Lumbricals, 13. Sartorius, 14. Extensor hallucis longus, 15. Extensor digitorum longus, 16. Soleus, 17. Tibialis anterior, 18. Gastrocnemius, 19. Quadriceps femoris, 20. Adductor magnus, 21. Adductor longus, 22. Flexor digitorum superficialis, 23. Tendon of palmaris longus, 24. Tendon of flexor carpi radialis, 25. Tendon of flexor carpi ulnaris, 26. Rectus abdominis, 27. Triceps brachii, 28. Brachialis, 29. Biceps brachii, 30. Deltoid, 31. Trapezius, 32. Orbicularis oris, 33. Orbicularis oculi, 34. Frontalis

Muscular System—Posterior View

Muscular System—Lateral View

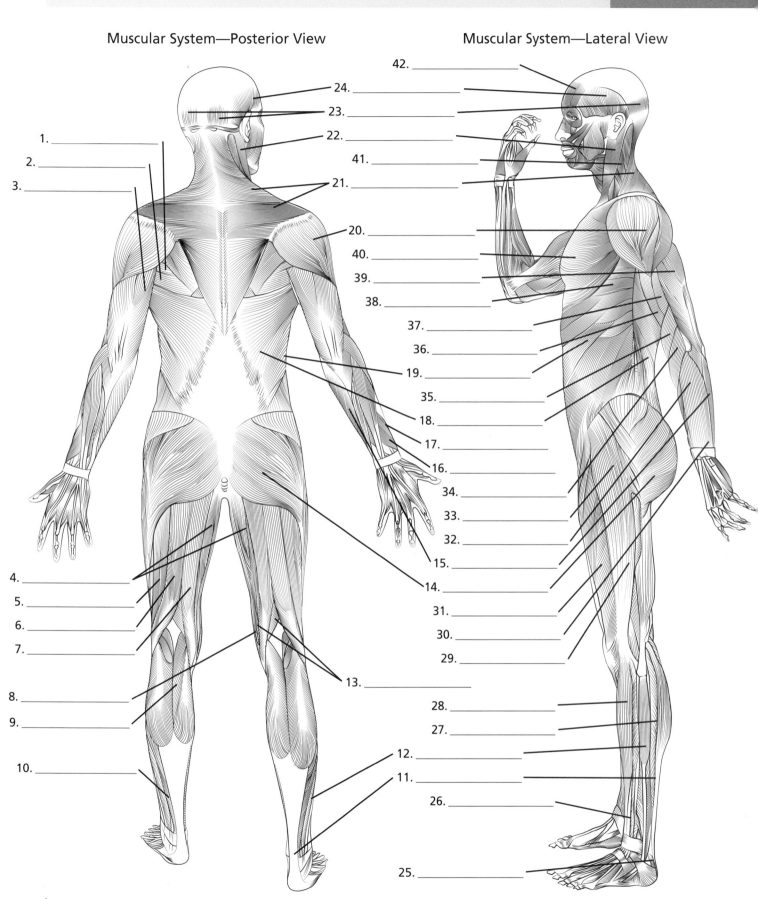

1. _____
2. _____
3. _____
4. _____
5. _____
6. _____
7. _____
8. _____
9. _____
10. _____

24. _____
23. _____
22. _____
21. _____
20. _____
19. _____
18. _____
17. _____
16. _____
15. _____
14. _____
13. _____
12. _____
11. _____

42. _____
41. _____
40. _____
39. _____
38. _____
37. _____
36. _____
35. _____
34. _____
33. _____
32. _____
31. _____
30. _____
29. _____
28. _____
27. _____
26. _____
25. _____

Answers

1. Teres minor, 2. Teres major, 3. Triceps brachii, 4. Adductor magnus, 5. Vastus lateralis, 6. Long head of biceps femoris, 7. Semitendinosus, 8. Gracilis, 9. Gastrocnemius, 10. Soleus, 11. Tendo calcaneus (Achilles tendon), 12. Fibularis (peroneus) longus, 13. Semimembranosus, 14. Gluteus maximus, 15. Flexor carpi ulnaris, 16. Extensor pollicis brevis, 17. Abductor pollicis longus, 18. Latissimus dorsi, 19. External abdominal oblique, 20. Deltoid, 21. Trapezius, 22. Sternocleidomastoid, 23. Occipitalis, 24. Temporalis, 25. Tendo calcaneus (Achilles tendon), 26. Extensor digitorum longus, 27. Lateral head of gastrocnemius, 28. Tibialis anterior, 29. Iliotibial tract, 30. Extensor carpi ulnaris, 31. Quadriceps femoris (vastus lateralis), 32. Extensor digitorum, 33. Tensor fasciae latae, 34. Extensor carpi radialis longus, 35. Brachioradialis, 36. Biceps brachii, 37. Brachialis, 38. Serratus anterior, 39. Lateral head of triceps brachii, 40. Pectoralis major, 41. Levator scapulae, 42. Frontalis

Muscles of the Head and Neck

Superficial and
Deep Muscles of
the Head and Neck—
Anterior View

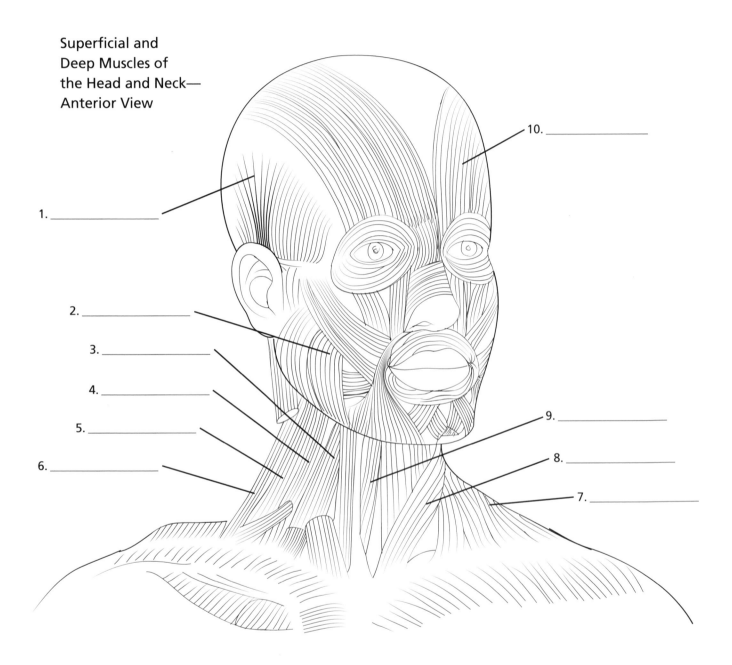

1. _____

2. _____

3. _____

4. _____

5. _____

6. _____

7. _____

8. _____

9. _____

10. _____

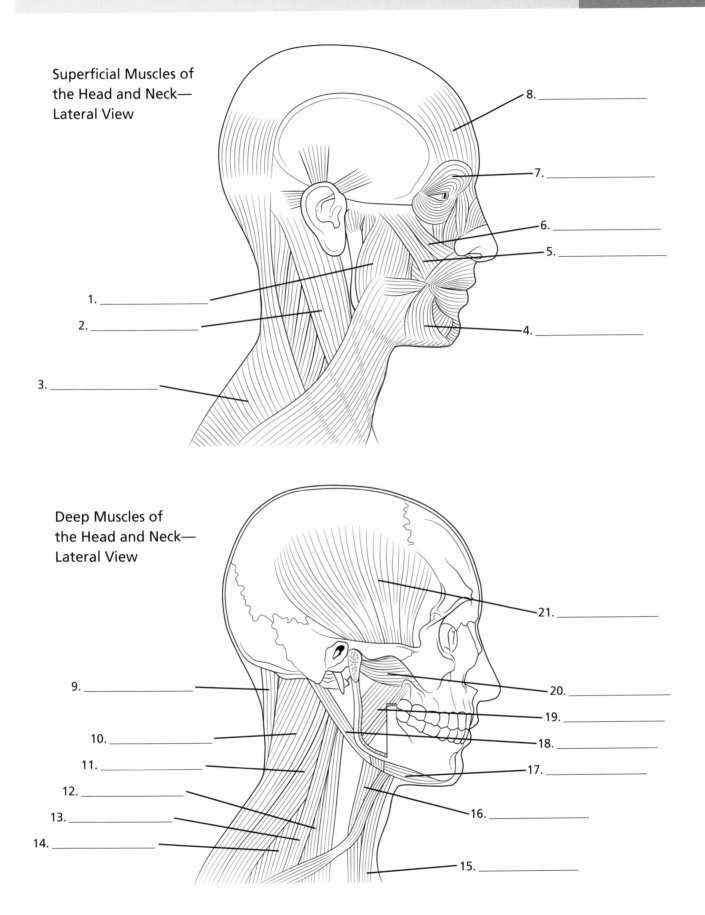

Superficial Muscles of the Head and Neck—Lateral View

8. _____
7. _____
6. _____
5. _____
1. _____
2. _____
4. _____
3. _____

Deep Muscles of the Head and Neck—Lateral View

21. _____
9. _____
20. _____
19. _____
10. _____
18. _____
11. _____
17. _____
12. _____
13. _____
16. _____
14. _____
15. _____

Answers

1. Masseter, 2. Sternocleidomastoid, 3. Trapezius, 4. Depressor anguli oris, 5. Zygomaticus minor, 6. Zygomaticus major, 7. Orbicularis oculi, 8. Frontalis, 9. Semispinalis capitis, 10. Splenius capitis, 11. Levator scapulae, 12. Scalenus anterior, 13. Scalenus medius, 14. Scalenus posterior, 15. Sternohyoid, 16. Thyrohyoid, 17. Digastric (anterior belly), 18. Digastric (posterior belly), 19. Medial pterygoid, 20. Lateral pterygoid, 21. Temporalis

Muscles of the Back

Superficial Muscles
of the Back—
Posterior View

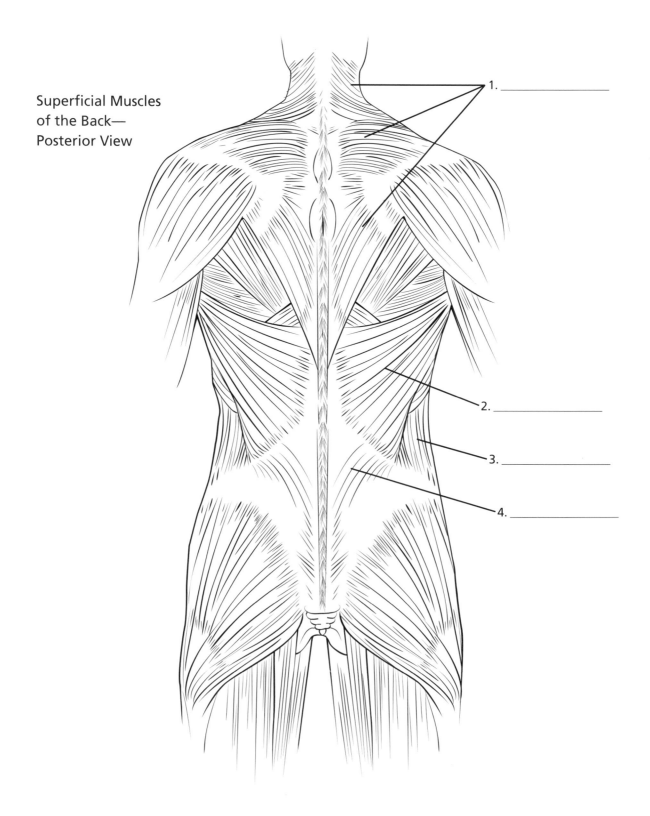

1. _____

2. _____

3. _____

4. _____

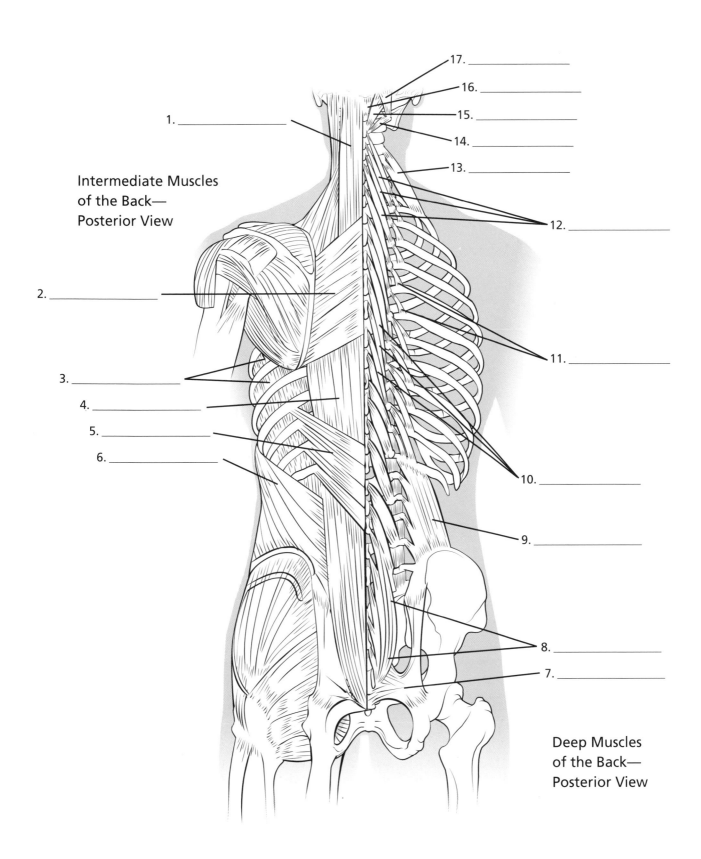

1. _____

**Intermediate Muscles
of the Back—
Posterior View**

2. _____

3. _____

4. _____

5. _____

6. _____

17. _____

16. _____

15. _____

14. _____

13. _____

12. _____

11. _____

10. _____

9. _____

8. _____

7. _____

**Deep Muscles
of the Back—
Posterior View**

Answers

1. Semispinalis capitis, 2. Rhomboid major, 3. External intercostals, 4. Erector spinae, 5. Serratus posterior inferior, 6. Internal oblique, 7. Sacrotuberous ligament, 8. Multifidus, 9. Quadratus lumborum, 10. Semispinalis thoracis, 11. Levatores costarum, 12. Semispinalis cervicis, 13. Scalenus posterior, 14. Obliquus capitis inferior, 15. Rectus capitis posterior major, 16. Rectus capitis posterior minor, 17. Obliquus capitis superior

Muscles of the Thorax and Abdomen

Superficial and Deep Muscles of the Thorax and Abdomen— Anterior View

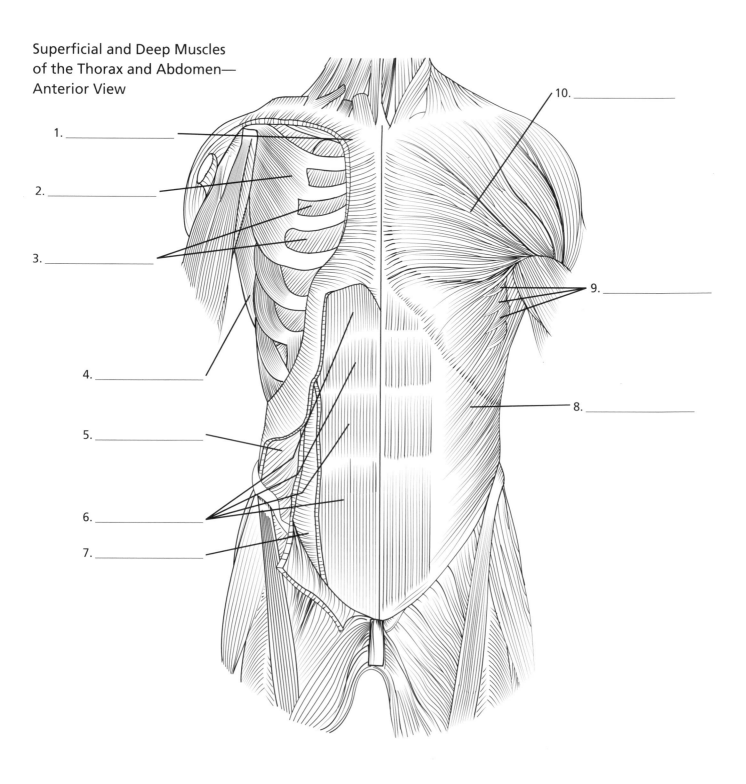

1. _____

2. _____

3. _____

4. _____

5. _____

6. _____

7. _____

8. _____

9. _____

10. _____

Answers

Muscles of the Shoulder

Superficial and Deep
Muscles of the Shoulder—
Posterior View

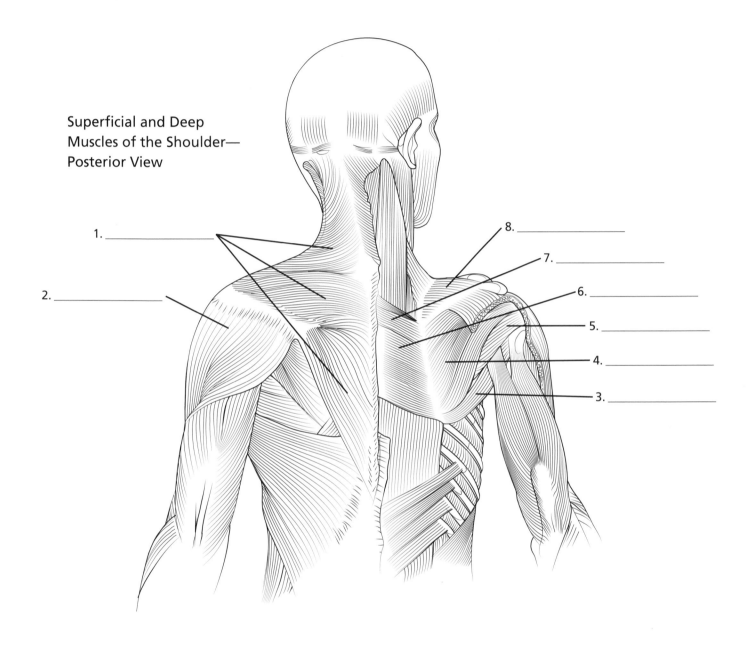

1. _____

2. _____

8. _____

7. _____

6. _____

5. _____

4. _____

3. _____

Answers

Muscles of the Shoulder

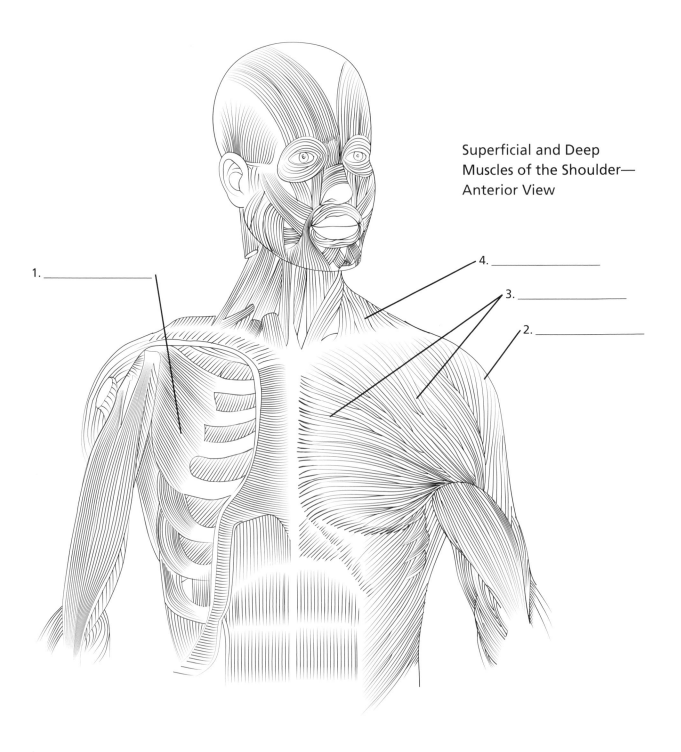

Superficial and Deep
Muscles of the Shoulder—
Anterior View

1. _____

4. _____

3. _____

2. _____

1. Pectoralis minor, 2. Deltoid, 3. Pectoralis major, 4. Trapezius

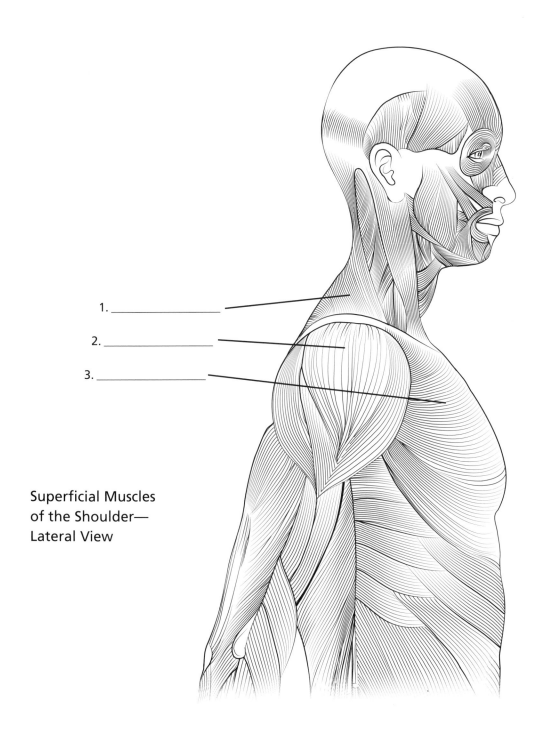

1. _____

2. _____

3. _____

Superficial Muscles
of the Shoulder—
Lateral View

Answers

Muscles of the Upper Limb

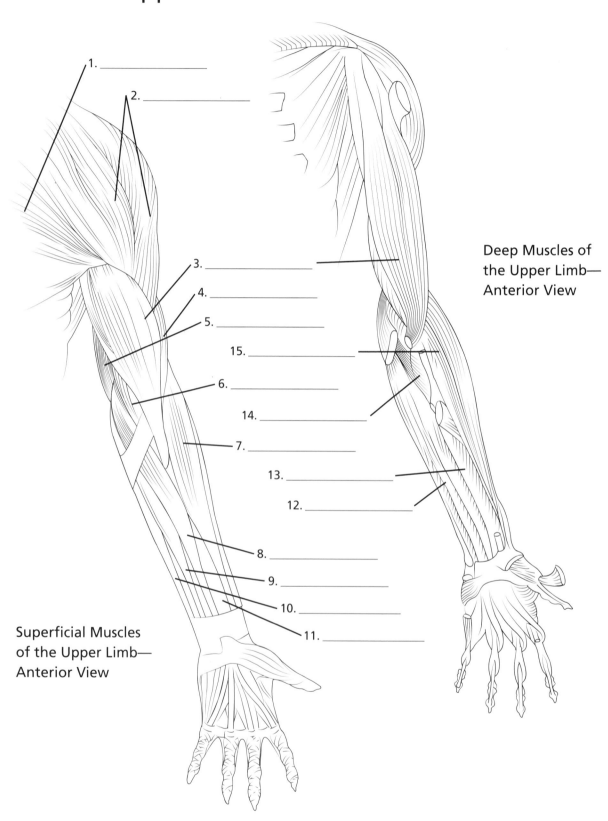

1. _____

2. _____

Deep Muscles of
the Upper Limb—
Anterior View

3. _____

4. _____

5. _____

15. _____

6. _____

14. _____

7. _____

13. _____

12. _____

8. _____

9. _____

10. _____

11. _____

Superficial Muscles
of the Upper Limb—
Anterior View

Answers

1. Pectoralis major, 2. Deltoid, 3. Biceps brachii, 4. Brachialis, 5. Triceps brachii, 6. Pronator teres, 7. Brachioradialis, 8. Tendon of flexor carpi radialis, 9. Tendon of palmaris longus, 10. Tendon of flexor carpi ulnaris, 11. Flexor digitorum superficialis, 12. Flexor digitorum profundus, 13. Flexor pollicis longus, 14. Pronator teres, 15. Extensor carpi radialis longus

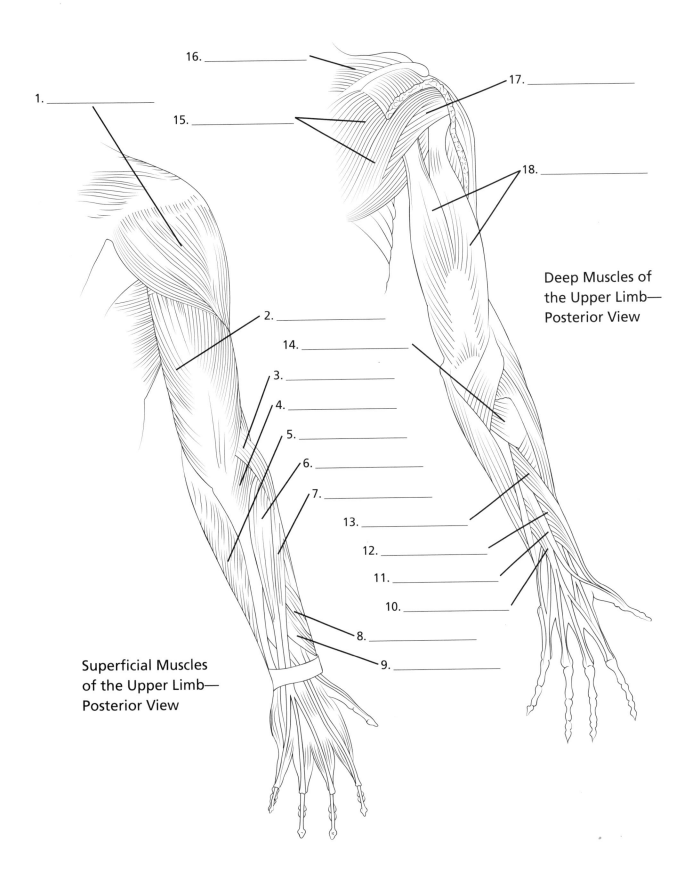

16. _____

1. _____

15. _____

17. _____

18. _____

Deep Muscles of the Upper Limb—Posterior View

2. _____

14. _____

3. _____

4. _____

5. _____

6. _____

7. _____

13. _____

12. _____

11. _____

10. _____

8. _____

9. _____

Superficial Muscles of the Upper Limb—Posterior View

Answers

Muscles of the Upper Limb

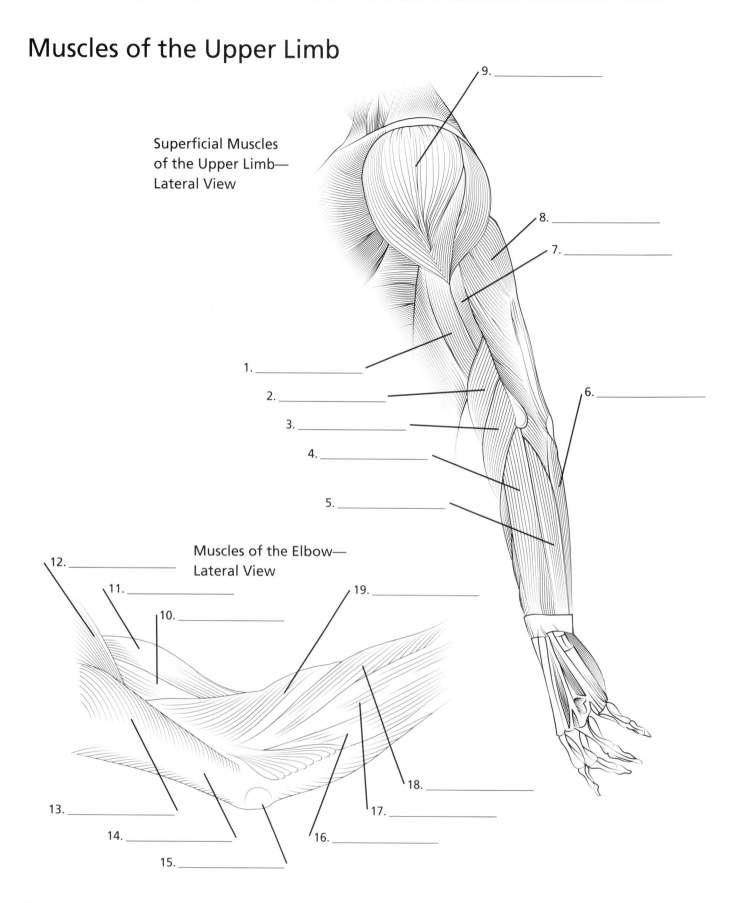

Superficial Muscles
of the Upper Limb—
Lateral View

9. _____

8. _____

7. _____

6. _____

1. _____

2. _____

3. _____

4. _____

5. _____

Muscles of the Elbow—
Lateral View

12. _____

11. _____

10. _____

19. _____

13. _____

14. _____

15. _____

16. _____

17. _____

18. _____

Answers

1. Biceps brachii, 2. Brachioradialis, 3. Extensor carpi radialis longus, 4. Extensor digitorum, 5. Extensor carpi ulnaris, 6. Flexor carpi ulnaris, 7. Brachialis, 8. Lateral head of triceps brachii, 9. Deltoid, 10. Brachialis, 11. Biceps brachii, 12. Deltoid, 13. Triceps brachii, 14. Tendon of triceps brachii, 15. Olecranon, 16. Extensor carpi ulnaris, 17. Extensor digitorum, 18. Extensor carpi radialis longus, 19. Brachioradialis

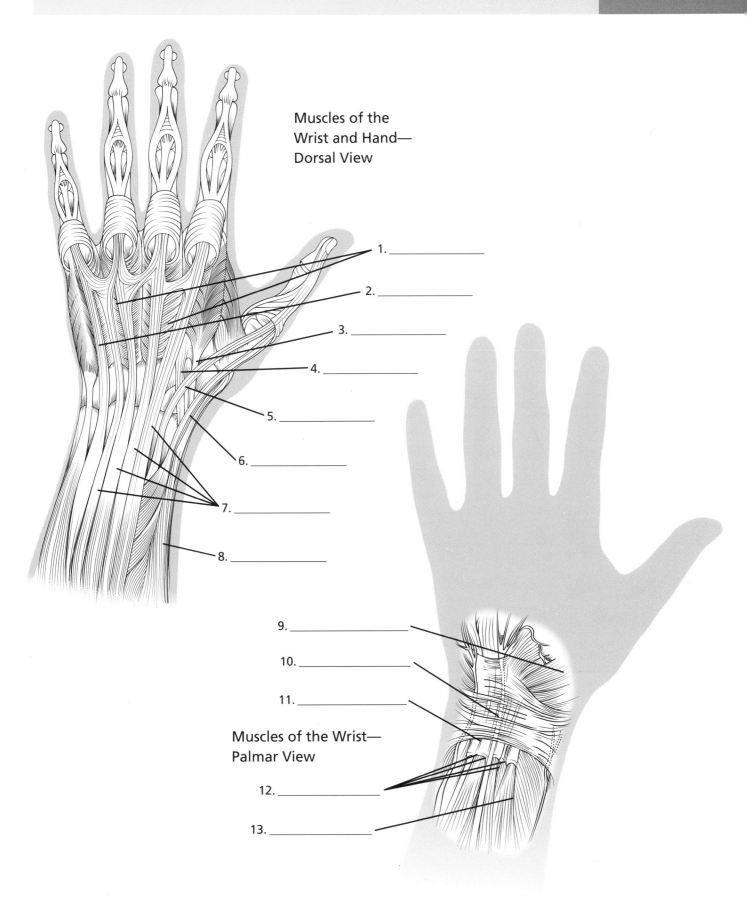

Muscles of the
Wrist and Hand—
Dorsal View

1. _____
2. _____
3. _____
4. _____
5. _____
6. _____
7. _____
8. _____

9. _____
10. _____
11. _____

Muscles of the Wrist—
Palmar View

12. _____
13. _____

Answers

Muscles of the Lower Limb

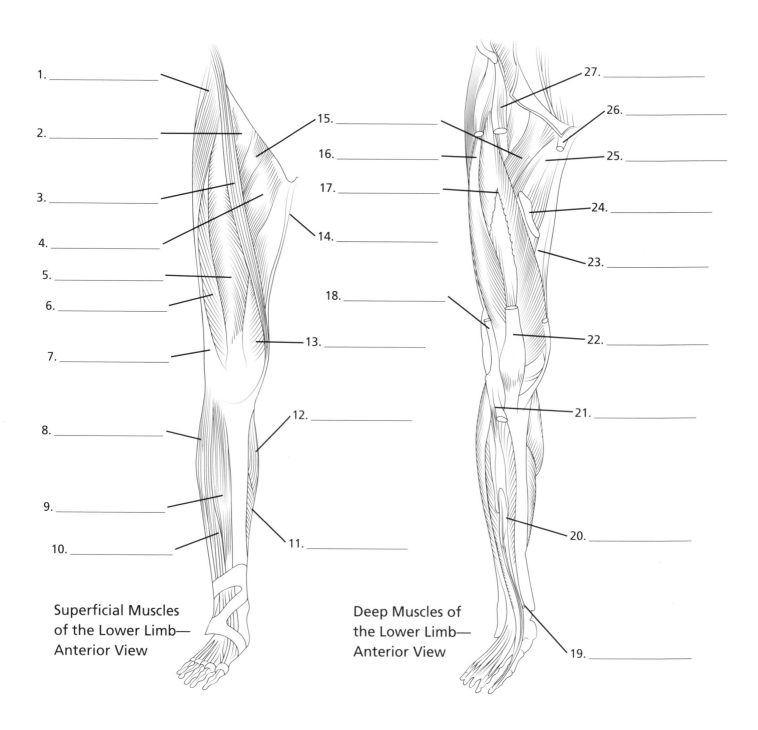

1. _____
2. _____
3. _____
4. _____
5. _____
6. _____
7. _____
8. _____
9. _____
10. _____

15. _____
16. _____
17. _____
14. _____
18. _____
13. _____
12. _____
11. _____

27. _____
26. _____
25. _____
24. _____
23. _____
22. _____
21. _____
20. _____
19. _____

Superficial Muscles
of the Lower Limb—
Anterior View

Deep Muscles of
the Lower Limb—
Anterior View

Answers

1. Tensor fasciae latae, 2. Iliopsoas, 3. Sartorius, 4. Adductor longus, 5. Rectus femoris, 6. Vastus lateralis, 7. Iliotibial tract, 8. Fibularis (peroneus) longus, 9. Tibialis anterior, 10. Extensor digitorum longus, 11. Soleus, 12. Gastrocnemius, 13. Vastus medialis, 14. Gracilis, 15. Pectineus, 16. Vastus lateralis, 17. Vastus intermedius, 18. Iliotibial tract (cut), 19. Tibialis anterior (cut), 20. Extensor hallucis longus, 21. Tibialis anterior (cut), 22. Rectus femoris (cut), 23. Adductor magnus, 24. Adductor longus (cut), 25. Adductor brevis, 26. Adductor longus (cut), 27. Sartorius (cut)

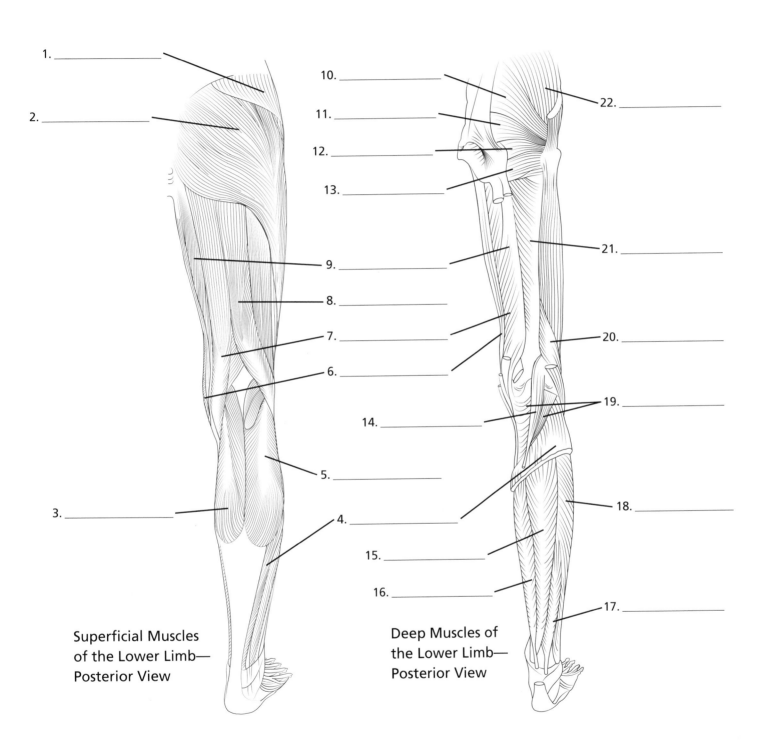

1. _____

2. _____

3. _____

9. _____

8. _____

7. _____

6. _____

5. _____

4. _____

**Superficial Muscles
of the Lower Limb—
Posterior View**

10. _____

11. _____

12. _____

13. _____

22. _____

21. _____

20. _____

19. _____

18. _____

17. _____

14. _____

15. _____

16. _____

**Deep Muscles of
the Lower Limb—
Posterior View**

Answers

1. Gluteus medius, 2. Gluteus maximus, 3. Medial head of gastrocnemius, 4. Soleus, 5. Lateral head of gastrocnemius, 6. Gracilis, 7. Semitendinosus, 8. Biceps femoris, 9. Adductor magnus, 10. Piriformis, 11. Superior gemellus, 12. Inferior gemellus, 13. Quadratus femoris, 14. Plantaris, 15. Tibialis posterior, 16. Flexor digitorum longus, 17. Flexor hallucis longus, 18. Fibularis (peroneus) longus, 19. Popliteus, 20. Short head of biceps femoris, 21. Adductor part of adductor magnus, 22. Gluteus minimus

Muscles of the Lower Limb

Superficial Muscles
of the Lower Limb—
Lateral View

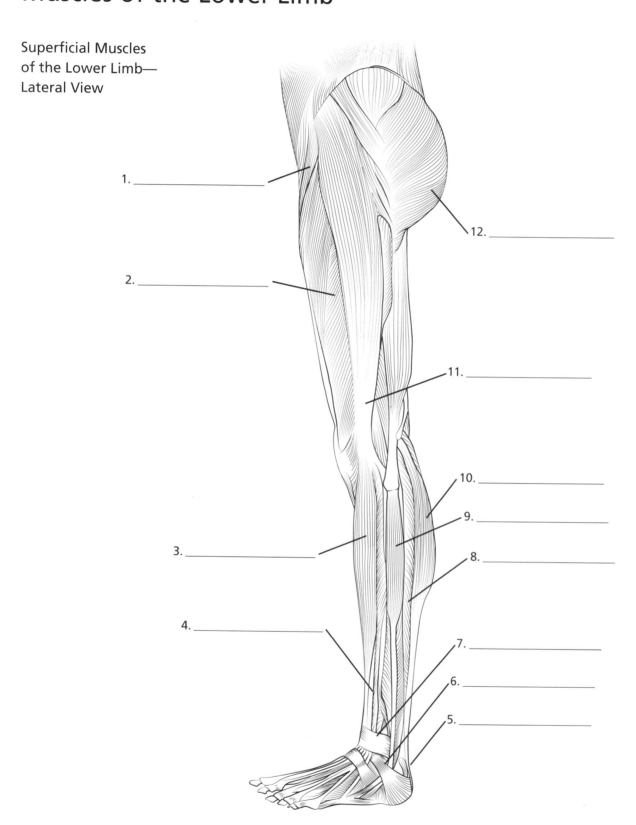

1._____

2._____

3._____

4._____

12._____

11._____

10._____

9._____

8._____

7._____

6._____

5._____

Answers

1. Sartorius, 2. Quadriceps femoris (vastus lateralis), 3. Tibialis anterior, 4. Extensor digitorum longus, 5. Tendo calcaneus (Achilles tendon), 6. Inferior extensor retinaculum, 7. Superior extensor retinaculum, 8. Soleus, 9. Fibularis (peroneus) longus, 10. Lateral head of gastrocnemius, 11. Iliotibial tract, 12. Gluteus maximus

1. _____
2. _____
3. _____
4. _____
5. _____
6. _____
7. _____

**Muscles of the Foot—
Lateral View**

8. _____
9. _____
10. _____
11. _____

15. _____
14. _____
13. _____
12. _____

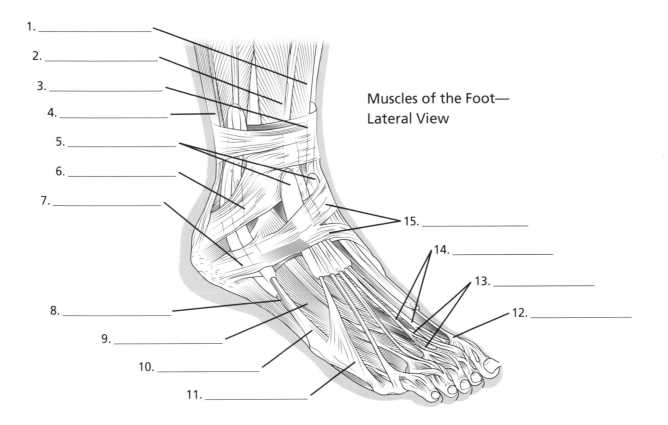

16. _____
17. _____
18. _____
19. _____
20. _____
21. _____
22. _____

**Muscles of the Foot—
Posteromedial View**

23. _____

28. _____
27. _____
26. _____
25. _____
24. _____

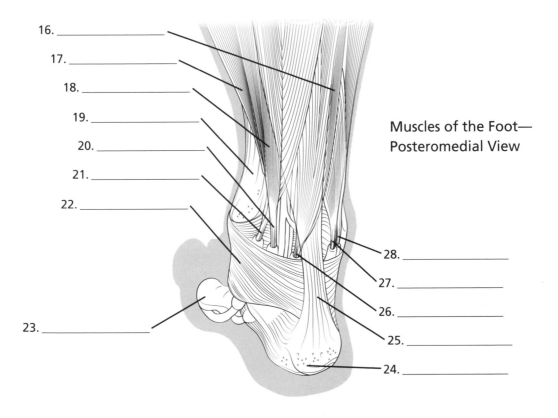

Answers

Muscle Types

1. _____

2. _____

3. _____

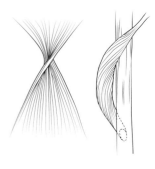

4. _____

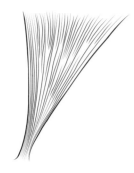

5. _____

6. _____

7. _____

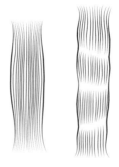

8. _____

9. _____

10. _____

11. _____

12. _____

13. _____

14. _____

15. _____

16. _____

Answers

Articulations

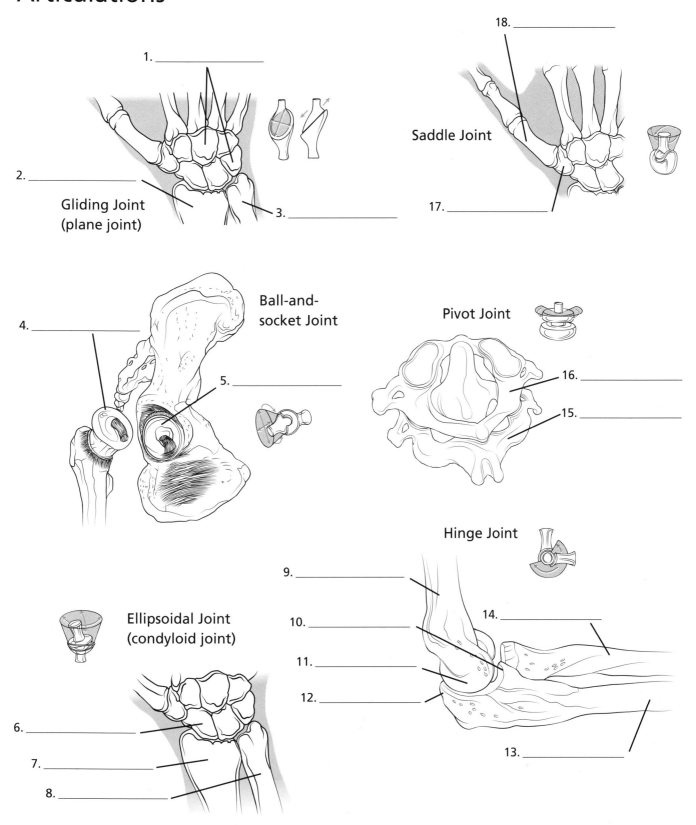

1. _____

2. _____

Gliding Joint
(plane joint)

3. _____

18. _____

Saddle Joint

17. _____

4. _____

Ball-and-
socket Joint

5. _____

Pivot Joint

16. _____

15. _____

Ellipsoidal Joint
(condyloid joint)

6. _____

7. _____

8. _____

Hinge Joint

9. _____

10. _____

11. _____

12. _____

14. _____

13. _____

Answers

1. Carpal bones, 2. Radius, 3. Ulna, 4. Head of femur (ball), 5. Acetabulum (socket), 6. Scaphoid bone, 7. Radius, 8. Ulna, 9. Humerus, 10. Coronoid process of ulna, 11. Trochlea (of humerus), 12. Olecranon, 13. Ulna, 14. Radius, 15. Axis, 16. Atlas, 17. Trapezium bone, 18. Metacarpal bone of thumb

Skeletal System

Skeletal System—
Anterior View

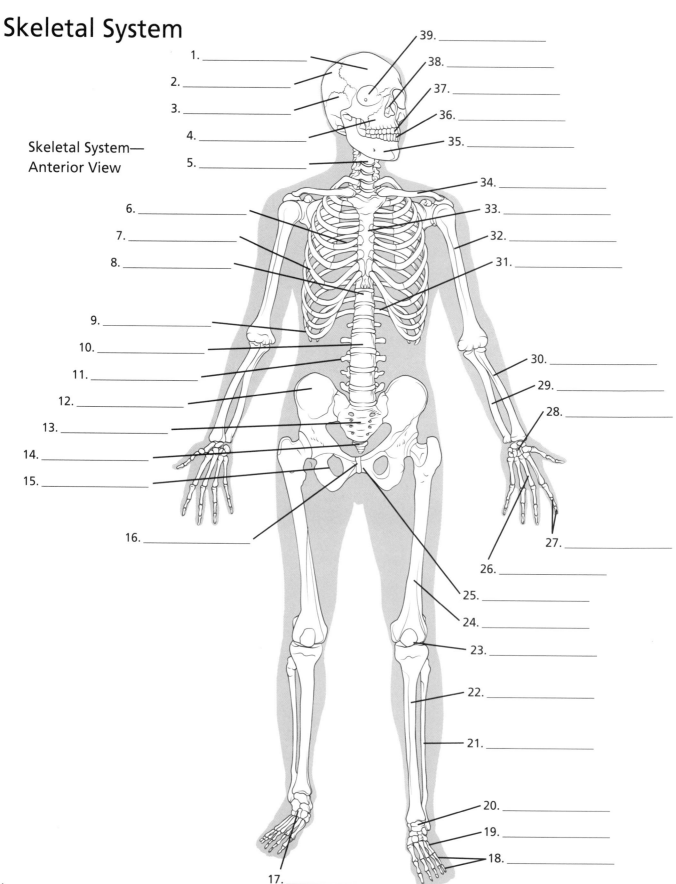

1. _____
2. _____
3. _____
4. _____
5. _____
6. _____
7. _____
8. _____
9. _____
10. _____
11. _____
12. _____
13. _____
14. _____
15. _____
16. _____
17. _____

39. _____
38. _____
37. _____
36. _____
35. _____
34. _____
33. _____
32. _____
31. _____
30. _____
29. _____
28. _____
27. _____
26. _____
25. _____
24. _____
23. _____
22. _____
21. _____
20. _____
19. _____
18. _____

Answers

1. Frontal bone, 2. Parietal bone, 3. Temporal bone, 4. Maxilla, 5. Cervical vertebra, 6. Costal cartilage, 7. True rib, 8. Thoracic vertebra, 9. False rib, 10. Lumbar vertebra, 11. Transverse process, 12. Ilium, 13. Sacrum, 14. Coccyx, 15. Ischium, 16. Pubic symphysis, 17. Tarsal bones, 18. Phalanges, 19. Metatarsal bones, 20. Talus, 21. Fibula, 22. Tibia, 23. Patella, 24. Femur, 25. Pubis, 26. Metacarpal bones, 27. Phalanges, 28. Carpal bones, 29. Ulna, 30. Radius, 31. Twelfth rib (floating rib), 32. Humerus, 33. Sternum, 34. Clavicle, 35. Mandible, 36. Lower teeth, 37. Upper teeth, 38. Anterior nasal (piriform) aperture, 39. Orbit

Skeletal System—Posterior View

Skeletal System—Lateral View

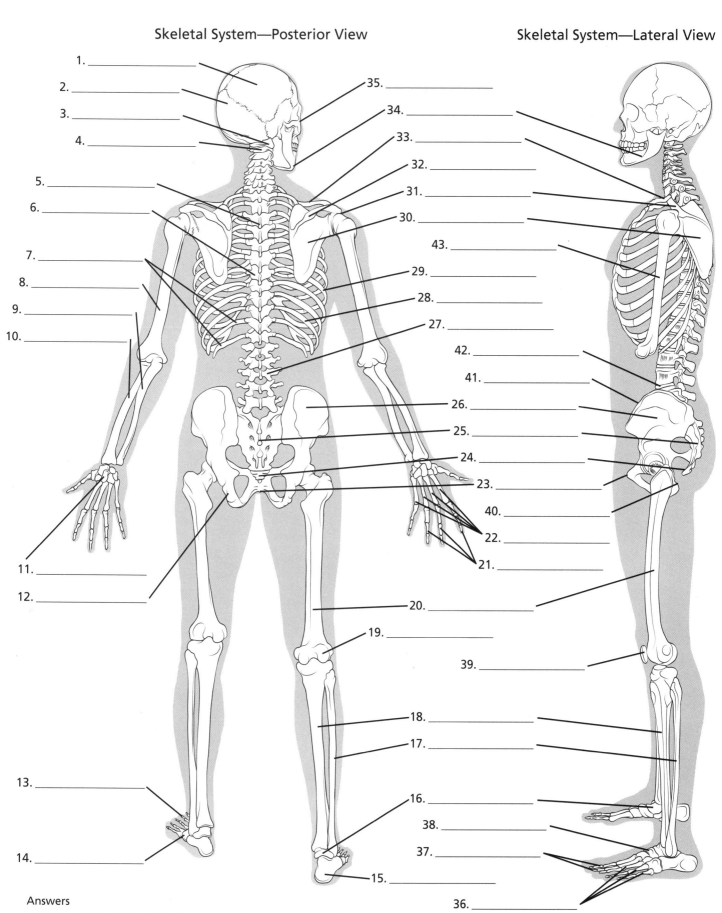

1. _____
2. _____
3. _____
4. _____
5. _____
6. _____
7. _____
8. _____
9. _____
10. _____
11. _____
12. _____
13. _____
14. _____
15. _____
16. _____
17. _____
18. _____
19. _____
20. _____
21. _____
22. _____
23. _____
24. _____
25. _____
26. _____
27. _____
28. _____
29. _____
30. _____
31. _____
32. _____
33. _____
34. _____
35. _____
36. _____
37. _____
38. _____
39. _____
40. _____
41. _____
42. _____
43. _____

Answers

Vertebral Column

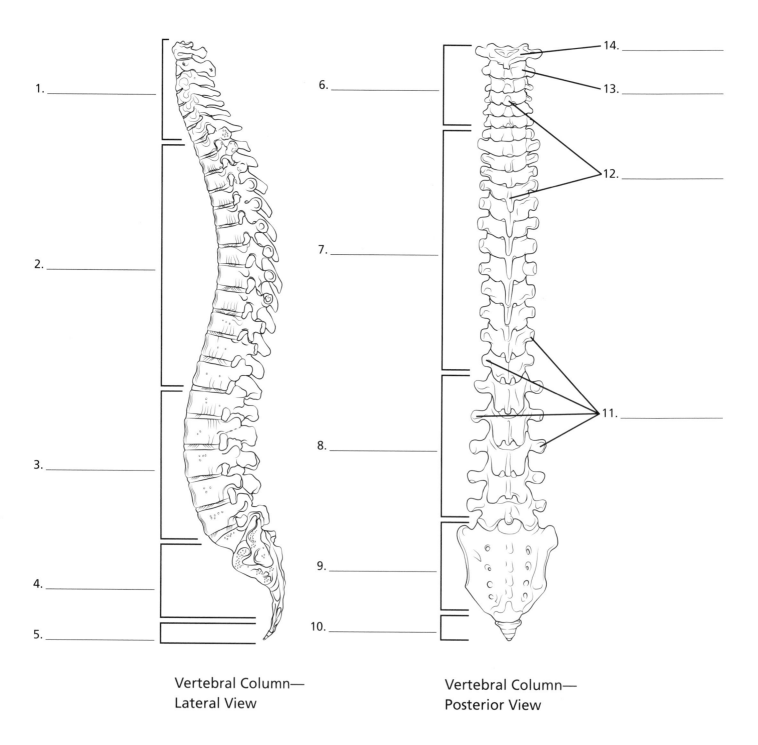

1. _____

2. _____

3. _____

4. _____

5. _____

6. _____

7. _____

8. _____

9. _____

10. _____

11. _____

12. _____

13. _____

14. _____

Vertebral Column—
Lateral View

Vertebral Column—
Posterior View

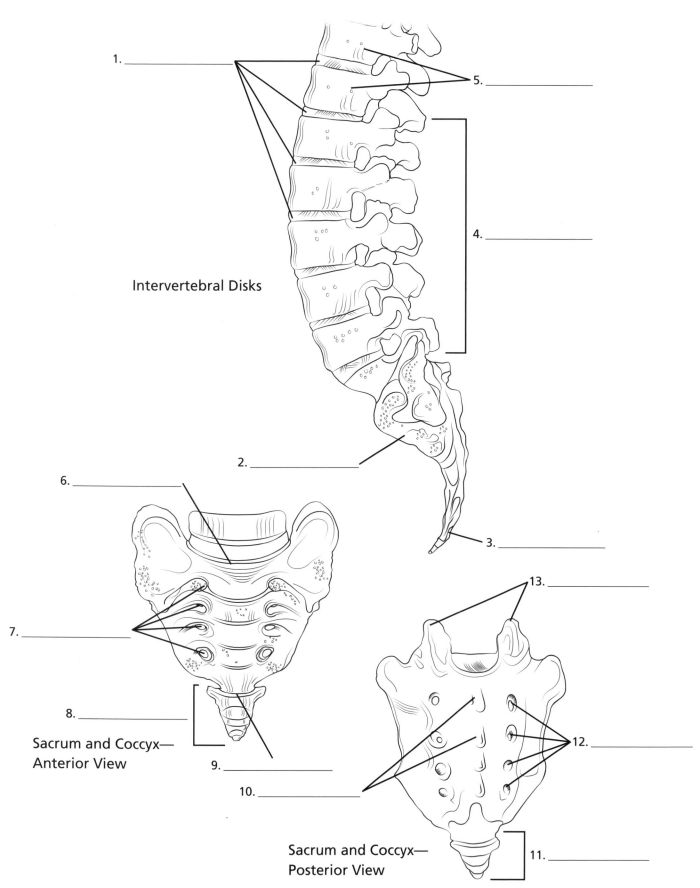

Intervertebral Disks

1. _____

5. _____

4. _____

2. _____

3. _____

6. _____

7. _____

8. _____

Sacrum and Coccyx—
Anterior View

9. _____

10. _____

13. _____

12. _____

11. _____

Sacrum and Coccyx—
Posterior View

Answers

1. Intervertebral disks, 2. Sacrum, 3. Coccyx, 4. Lumbar vertebrae, 5. Thoracic vertebrae, 6. Sacral promontory, 7. Pelvic sacral foramina, 8. Coccyx, 9. Sacrococcygeal joint, 10. Median sacral crest and spinous tubercles, 11. Coccyx, 12. Posterior sacral foramina, 13. Superior articular processes (facets)

Bones of the Upper Limb

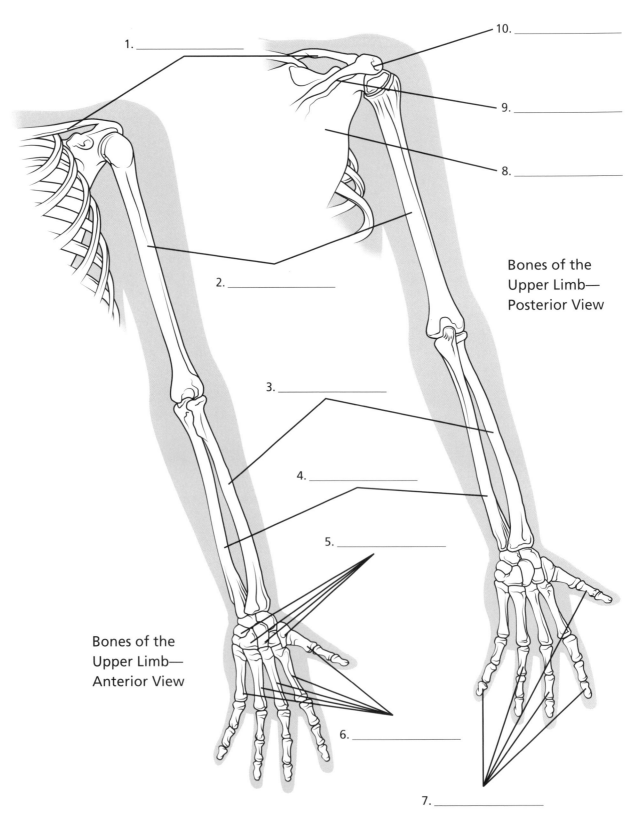

1. _____

2. _____

3. _____

4. _____

5. _____

6. _____

7. _____

8. _____

9. _____

10. _____

Bones of the
Upper Limb—
Posterior View

Bones of the
Upper Limb—
Anterior View

Answers

1. Clavicle, 2. Humerus, 3. Radius, 4. Ulna, 5. Carpal bones, 6. Metacarpal bones, 7. Phalanges, 8. Scapula, 9. Spine of scapula, 10. Acromion

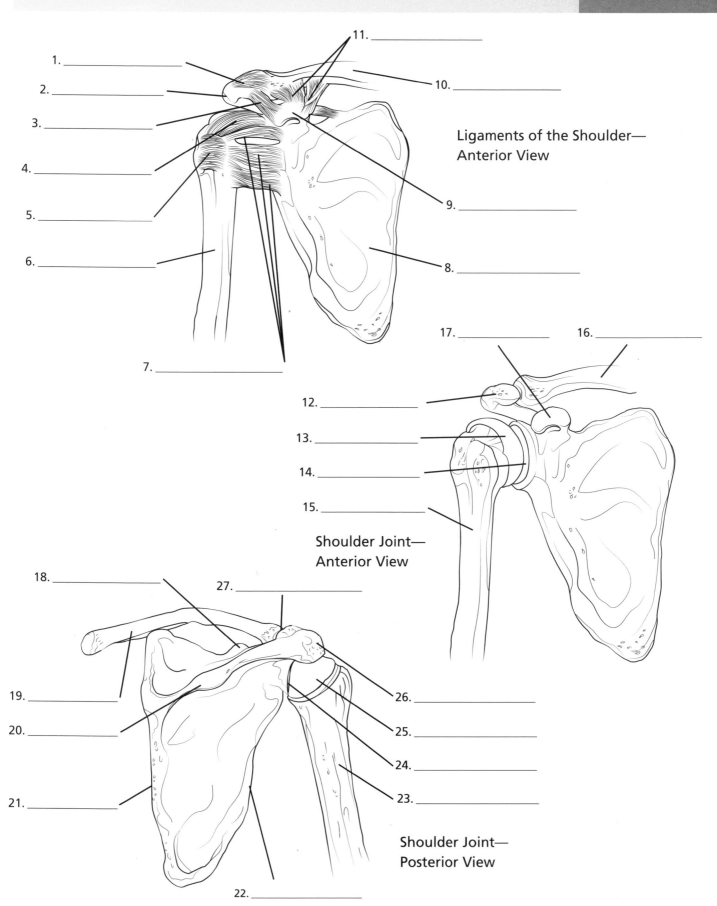

1. _____

2. _____

3. _____

4. _____

5. _____

6. _____

7. _____

11. _____

10. _____

9. _____

8. _____

**Ligaments of the Shoulder—
Anterior View**

17. _____

16. _____

12. _____

13. _____

14. _____

15. _____

**Shoulder Joint—
Anterior View**

18. _____

27. _____

19. _____

20. _____

21. _____

26. _____

25. _____

24. _____

23. _____

22. _____

**Shoulder Joint—
Posterior View**

Answers

Bones of the Upper Limb

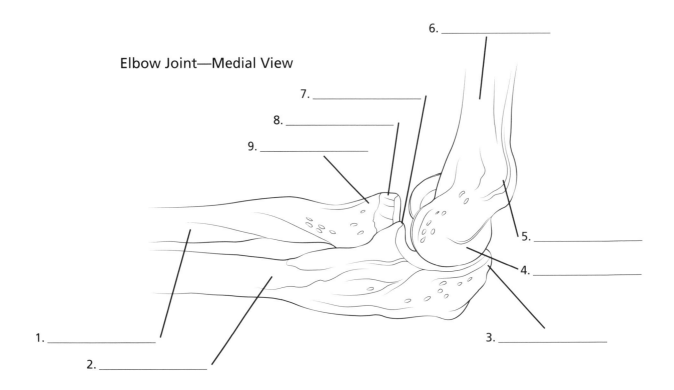

Elbow Joint—Medial View

6. _____

7. _____

8. _____

9. _____

5. _____

4. _____

1. _____

2. _____

3. _____

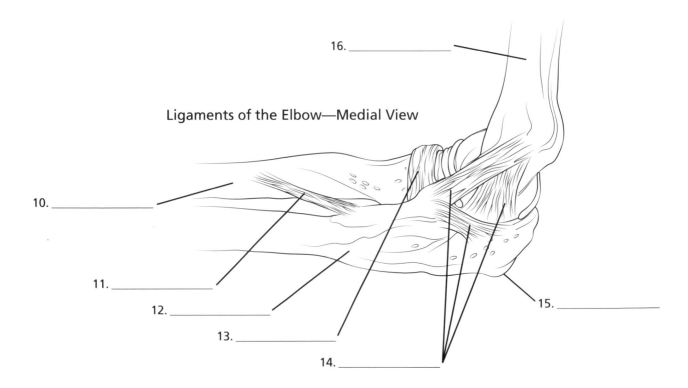

16. _____

Ligaments of the Elbow—Medial View

10. _____

11. _____

12. _____

13. _____

14. _____

15. _____

Answers

1. Radius, 2. Ulna, 3. Olecranon, 4. Trochlea of humerus, 5. Medial epicondyle of humerus, 6. Humerus, 7. Coronoid process of ulna, 8. Head of radius, 9. Neck of radius, 10. Radius, 11. Oblique cord, 12. Ulna, 13. Annular ligament of radius, 14. Ulnar collateral ligaments, 15. Olecranon, 16. Humerus

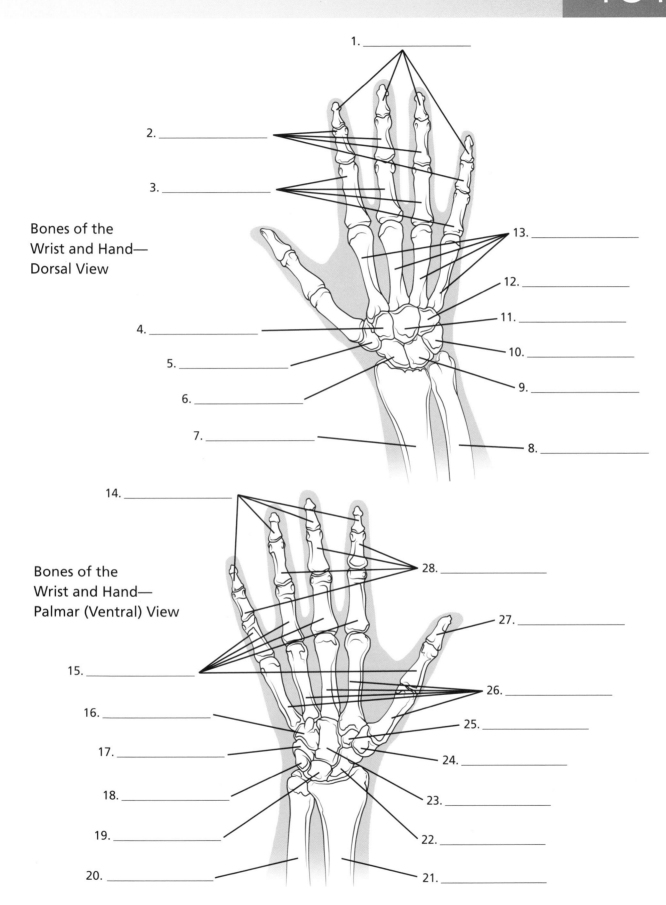

Bones of the
Wrist and Hand—
Dorsal View

1. _____

2. _____

3. _____

4. _____

5. _____

6. _____

7. _____

8. _____

9. _____

10. _____

11. _____

12. _____

13. _____

Bones of the
Wrist and Hand—
Palmar (Ventral) View

14. _____

15. _____

16. _____

17. _____

18. _____

19. _____

20. _____

21. _____

22. _____

23. _____

24. _____

25. _____

26. _____

27. _____

28. _____

Answers

1. Distal phalanges, 2. Middle phalanges, 3. Proximal phalanges, 4. Trapezoid, 5. Trapezium, 6. Scaphoid, 7. Radius, 8. Ulna, 9. Lunate, 10. Triquetrum, 11. Capitate, 12. Hamate, 13. Metacarpal bones, 14. Distal phalanges, 15. Proximal phalanges, 16. Hamate, 17. Triquetrum, 18. Pisiform, 19. Lunate, 20. Ulna, 21. Radius, 22. Scaphoid, 23. Capitate, 24. Trapezoid, 25. Trapezium, 26. Metacarpal bones, 27. Distal phalanx of thumb, 28. Middle phalanges

Bones of the Lower Limb

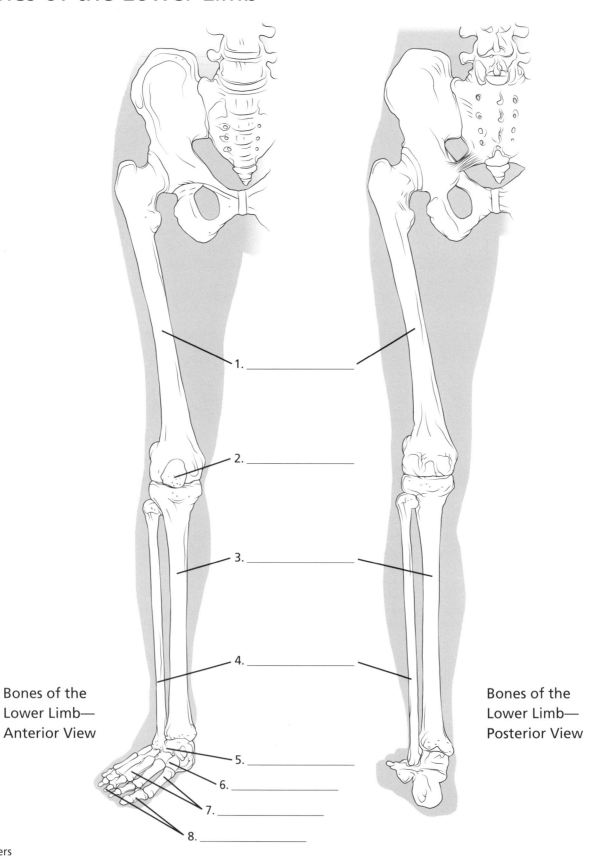

1. _____

2. _____

3. _____

Bones of the
Lower Limb—
Anterior View

4. _____

5. _____

6. _____

7. _____

8. _____

Bones of the
Lower Limb—
Posterior View

Answers

1. Femur, 2. Patella, 3. Tibia, 4. Fibula, 5. Talus, 6. Tarsal bones, 7. Metatarsal bones, 8. Phalanges

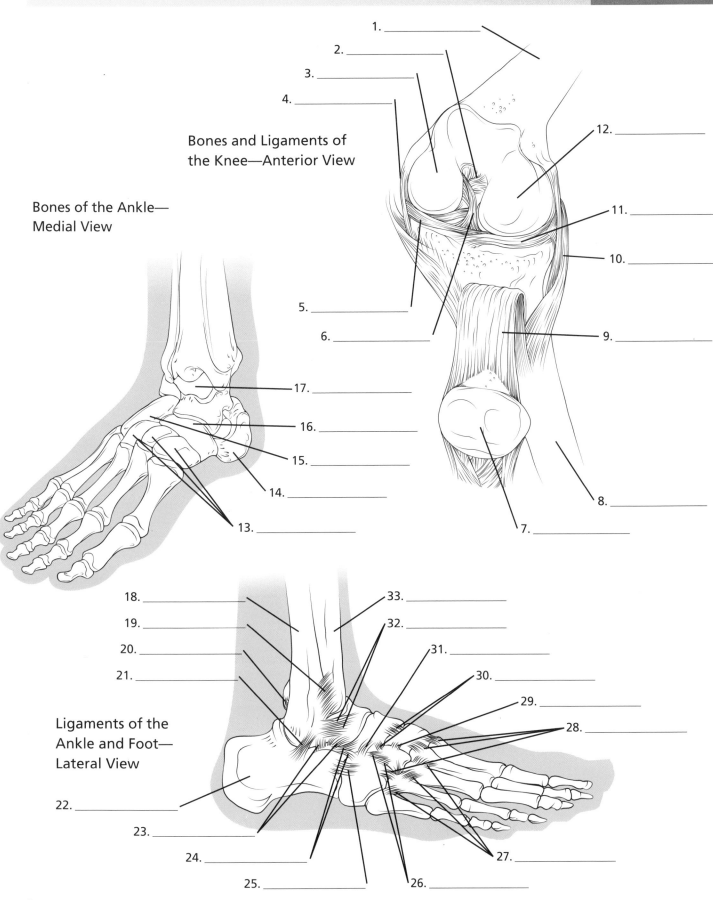

1. _____
2. _____
3. _____
4. _____

Bones and Ligaments of
the Knee—Anterior View

Bones of the Ankle—
Medial View

12. _____
11. _____
10. _____

5. _____
6. _____
9. _____

17. _____
16. _____
15. _____
14. _____
13. _____

8. _____
7. _____

18. _____
19. _____
20. _____
21. _____

33. _____
32. _____
31. _____
30. _____
29. _____
28. _____

Ligaments of the
Ankle and Foot—
Lateral View

22. _____
23. _____
24. _____
25. _____
26. _____
27. _____

Answers

Nerves of the Upper and Lower Limb

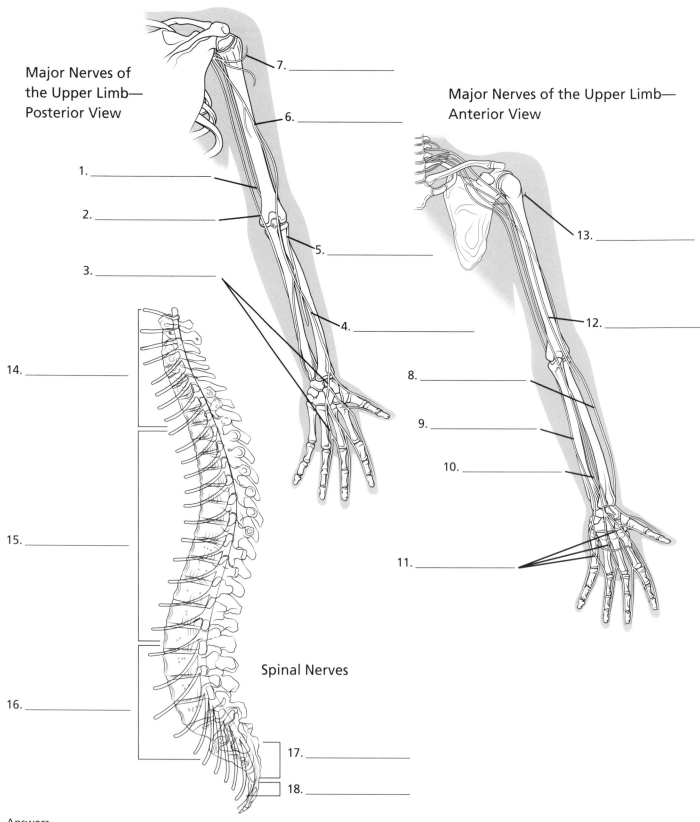

Major Nerves of the Upper Limb— Posterior View

7. _____

6. _____

1. _____

2. _____

5. _____

3. _____

Major Nerves of the Upper Limb— Anterior View

13. _____

12. _____

4. _____

8. _____

9. _____

10. _____

11. _____

14. _____

15. _____

16. _____

Spinal Nerves

17. _____

18. _____

Answers

Major Nerves of the Lower Limb—Posterior View

Major Nerves of the Lower Limb—Anterior View

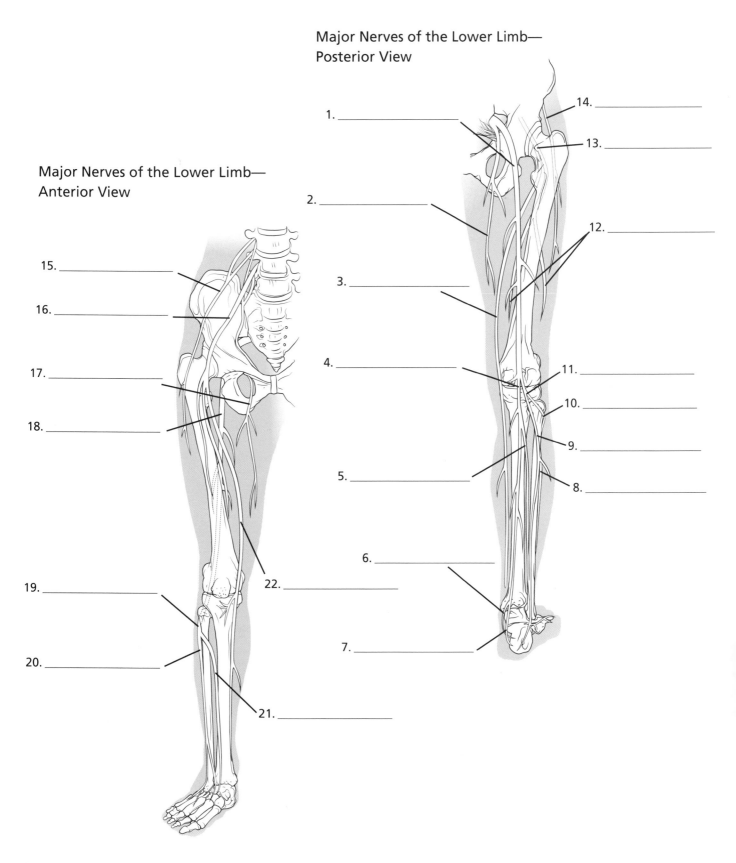

1. _____

2. _____

3. _____

4. _____

5. _____

6. _____

7. _____

8. _____

9. _____

10. _____

11. _____

12. _____

13. _____

14. _____

15. _____

16. _____

17. _____

18. _____

19. _____

20. _____

21. _____

22. _____

Answers

1. Sciatic nerve, 2. Posterior femoral cutaneous nerve, 3. Saphenous nerve, 4. Tibial nerve, 5. Medial sural cutaneous nerve, 6. Medial plantar nerve, 7. Lateral plantar nerve, 8. Lateral sural cutaneous nerve, 9. Deep fibular (peroneal) nerve, 10. Superficial fibular (peroneal) nerve, 11. Common fibular (peroneal) nerve, 12. Branches from femoral nerve, 13. Femoral nerve, 14. Lateral femoral cutaneous nerve, 15. Lateral femoral cutaneous nerve, 16. Femoral nerve, 17. Obturator nerve, 18. Sciatic nerve, 19. Common fibular (peroneal) nerve, 20. Superficial fibular (peroneal) nerve, 21. Deep fibular (peroneal) nerve, 22. Saphenous nerve

Reference

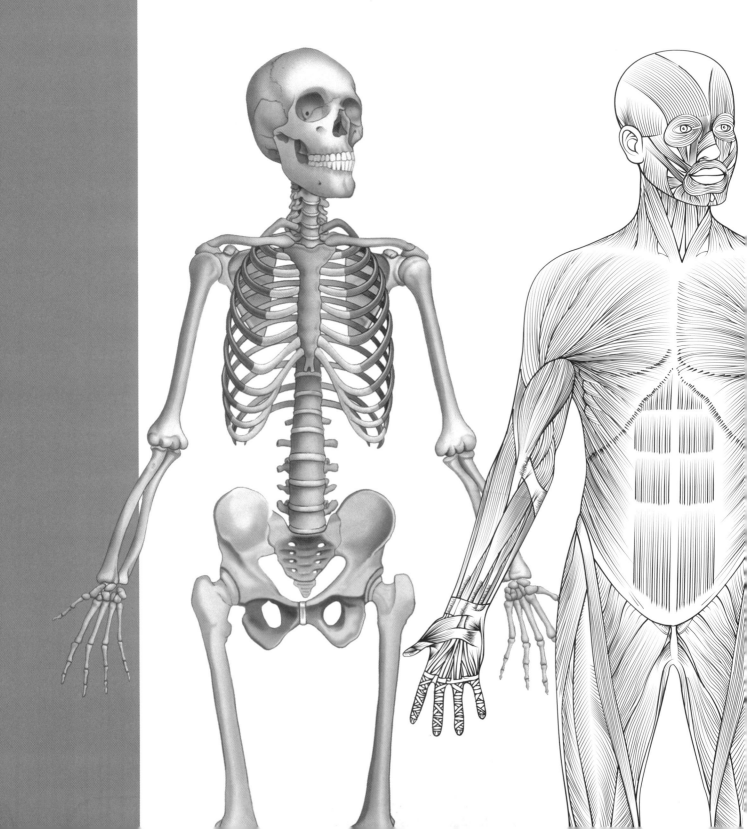

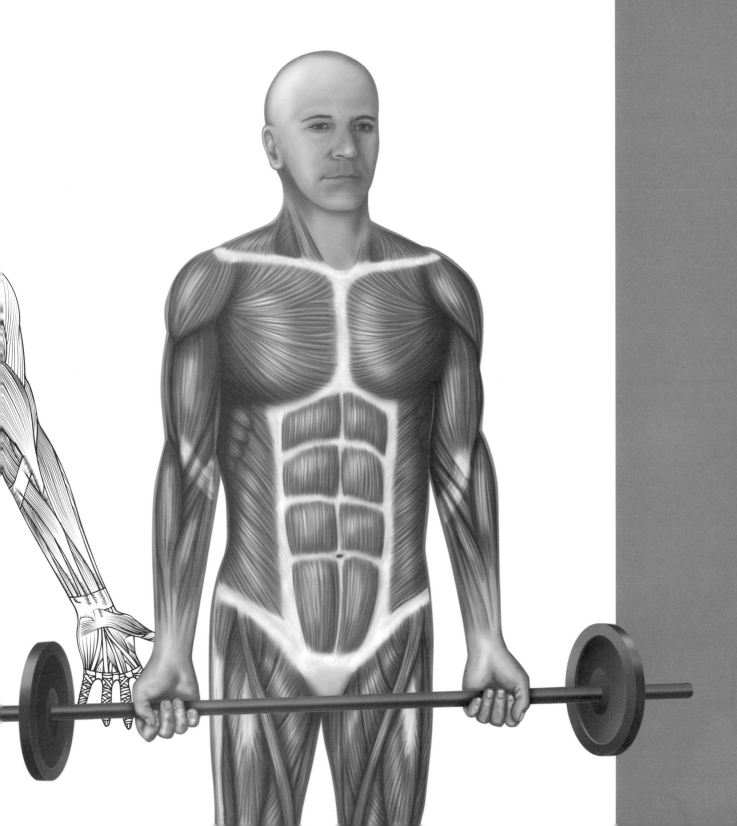

Glossary

Abduction Movement of a limb away from the midline of the body.

Accessory muscles *see* secondary muscles or movers.

Active muscles Muscles responsible for the main movement involved in an exercise. Also known as primary movers.

Adduction Movement of a limb toward the midline of the body.

Alternating grip Grasping the barbell with the palm of one hand facing away from the body, and the other palm facing toward the body.

Bilateral exercise An exercise using both limbs together.

Buttocks (muscles) This group includes the gluteus maximus on the surface, and the gluteus medius and minimus muscles beneath.

Cervical vertebrae The seven vertebrae of the neck. Together they form a curve that is concave to the back of the body.

Concentric contraction A muscle action where the muscle shortens as a result of contraction.

Concentric phase That part of an exercise movement where the muscle shortens as it contracts.

Core The trunk. Often used in reference to core stabilizing muscles, e.g., transverse abdominis, multifidus, obliques.

Eccentric contraction A muscle action where the muscle lengthens under tension, such as when lowering a weight against gravity in a controlled movement.

Eccentric control The smooth execution of an eccentric movement.

Eccentric overload The action whereby extra weight is resisted as the muscle lengthens while contracting, such as when lowering a barbell back to the starting position during the bicep curl. The eccentric action can handle more weight than the concentric (muscle shortening) action, so additional weight usually must be added to the barbell if the exerciser wishes to overload the eccentric phase.

Eccentric phase That part of an exercise movement where the muscle lengthens while under activation.

Extension The act of straightening a limb at a joint.

Extensor muscles A group of muscles that perform an extension movement.

Flexion The act of bending a limb at a joint.

Flexor muscles A group of muscles that perform a flexion movement.

Foot strike The impact of the foot on the ground during an exercise movement.

Hamstrings Muscles of the back of the thigh. The hamstring muscle group includes the semitendinosus, semimembranosus, and biceps femoris muscles.

Hyperextension Extension of a joint beyond the normal range of movement.

Internal rotation Rotating a limb toward the midline. Also called medial rotation.

Isometric contraction A muscle contraction where there is no change in the length of the muscle.

Isometrically active Activation of a muscle while staying the same length.

Kinetic chain The connection of all the parts of the body to one another, directly or indirectly. Moving one part of the body can affect the position and momentum of another part of the body.

Kyphosis A curvature of the vertebral column that is concave to the front of the body. It is normal in the thoracic spine.

Lordosis A curvature of the vertebral column that is concave to the back of the body. It is normal in the lumbar and cervical spine.

Lumbar vertebrae The five vertebrae between the bottom of the rib cage and the pelvis. They comprise the lower back.

Medial rotation *see* internal rotation.

Neutral spine The position of the vertebral column or spine, where there is the least stress on joints, ligaments, and disks. In the lumbar spine, this is a position of slight lordosis.

Posterior core Core stabilizers of the back, e.g., multifidus.

Posture During weight training, the correct position for the back is to maintain the normal lordosis curvature of the lumbar spine. Avoid flattening out or arching the back.

Primary mover The main muscle or muscles that produce a movement.

Progressive overload The progressive addition of resistive force to provide optimal training of a muscle.

Proprioceptive neuromuscular facilitation The role of sensory feedback from the stretch receptors in a muscle or its tendon to maintain muscle tone during contraction.

Quadriceps The quadriceps femoris muscle group on the front of the thigh includes the vastus lateralis, vastus intermedius, vastus medialis, and rectus femoris muscles.

Retraction Backward movement of the scapula toward the vertebral column.

Rotator cuff muscles A group of muscles (supraspinatus, subscapularis, infraspinatus, and teres minor) that arise from the scapula and insert onto the humerus to provide dynamic stability for the shoulder joint.

Scapula The bone commonly known as the shoulder blade.

Secondary muscles or movers Muscles that assist with a movement, but are not primary movers.

Spotter Someone who assists the exerciser to get into the correct starting position, ensures the barbell or dumbbell travels in the right direction, and assists if the exerciser struggles to control the weight during the exercise.

Stabilizers Muscles that are not involved in a movement, but help maintain body position. Important in the kinetic chain concept.

Super set Two exercises performed back to back without rest in between, to create a more intense workout.

Thoracic vertebrae The 12 vertebrae of the chest, comprising the upper and middle back—each is attached to a rib.

Unilateral exercise A single limb exercise.

Index

DISCARD

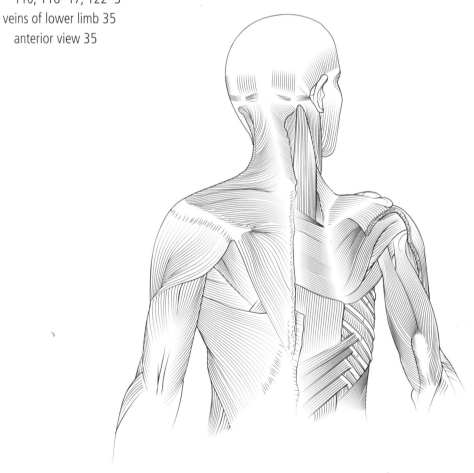